FOOD AND BEVERAGE CONSUMPTION AND HEALTH

CAPSAICIN

FOOD SOURCES, MEDICAL USES AND HEALTH IMPLICATIONS

FOOD AND BEVERAGE CONSUMPTION AND HEALTH

Additional books in this series can be found on Nova's website under the Series tab.

Additional e-books in this series can be found on Nova's website under the e-book tab.

FOOD AND BEVERAGE CONSUMPTION AND HEALTH

CAPSAICIN

FOOD SOURCES, MEDICAL USES AND HEALTH IMPLICATIONS

BRUCE GILLIAM
EDITOR

New York

Library of Congress Cataloging-in-Publication Data

ISBN: 978-1-61470-433-1

Library of Congress Control Number: 2014951170

Published by Nova Science Publishers, Inc. † New York

CONTENTS

PREFACE

Capsaicin, and several related compounds, are called capsaicinoids and are produced as secondary metabolites by chili peppers. Capsaicin is an alkaloid responsible for the spicy, burning taste of chili peppers. Capsaicin is a crystalline, lipophilic, colorless and odorless alkaloid. It is widely consumed in spicy foods throughout the world, particularly in South East Asia and Latin-America. This book discusses several medical uses of capsaicin as well as health implications and food sources.

Chapter 1 - The book chapter titled: "Creative Destruction in the Peripheral Nervous System: The action of capsaicin on primary sensory neurons", includes research on changes in the sensory component (SN) of the peripheral nervous system (PNS), which are the result of the neurotoxic actions of capsaicin. Capsaicin (8-methyl-N-vanillyl-6-nonenamid) is an alkaloid responsible for the spicy, burning taste of chili peppers. In mammals, it creates a burning sensation in the tissue when it comes in contact. Capsaicin works by the vanilloid receptor (TRPV1) and has neurotoxic properties. Intraperitoneal injection of capsaicin has become an appropriate tool to damage the sensory afferent fibers innervating the visceral organs, originating from the nodose ganglion (NG) and the dorsal root ganglia (DRG). Intraperitoneal injection of capsaicin caused the death of about half of the neurons in the studied ganglia.

A surprising, and very intriguing, fact is after about two months after the injury, the number of ganglion neurons was restored to the level observed in control animals. The above observation allowed us to test the hypothesis that the capsaicin-induced neuronal death in sensory ganglia activates a cascade of events leading to proliferation of ganglion cells and formation of new neurons, which restore the damaged peripheral sensory pathways. Further studies revealed the presence of nestin, a marker of stem cells in the satellite glia cells (SG), which proves their proliferative potential. SG, under the influence of events induced by capsaicin, enter the cell cycle, synthesize DNA (BrdU incorporation) and show immunoreactivity against Ki-67, an antigen present in the active phases of the cell cycle. Both the central axons from NG neurons forming synaptic connections in the medulla and the peripheral which innervate the upper gastrointestinal tract, are rebuilt after sixty days following the capsaicin administration. The authors used intraperitoneal administration of cholecystokinin (CCK), which is specifically acting via the vagus nerve to reduce appetite. The author's results showed that capsaicin significantly reduced the effect of CCK, which confirmed the damage to the peripheral axons of the vagus nerve. However, three hundred days after injury, the effect of CCK was restored. In addition, and very interesting, is the fact that at this time, the number and immunohistochemical characterization of neurons in the NG returned to that observed in control animals. In summary,the author's studies clearly show that capsaicin kills a significant population of viscerosensory neurons in PNS but also induces a cascade of events leading to an activation of endogenous multipotent cells in the sensory ganglia. These cells enter the cell cycle and divide and differentiate to form new glia and functional neurons. However, the molecular mechanism of the neurogenic action of capsaicin needs further investigation. Complete understanding of neurogenic capsaicin action and the ability to manipulate the population of endogenous ganglionic progenitors is a fundamental step towards the use of alternative reserve of stem cells to repair a damaged nervous system. This knowledge will help to develop methods to repair damaged CNS, arising as a result of an illness or accident.

Chapter 2 - Capsaicinoids are important for the food and pharmaceutical industries. For this reason, a number of researchers are engaged in improving their production by manipulating chili plant cultivation conditions, chemical and enzymatic synthesis processes or alternative methods such as cell or tissue culture. Hitherto, research has shown that capsaicinoids, and capsaicin in particular, have a wide variety of biological and physiological activities, providing them with different functions such as antioxidant, anticarcinogenic, anti-inflammatory, promoter of energy metabolism and suppressor of fat accumulation. However, the potential applications of these molecules are limited by the irritation caused by their pungency. This has driven the search and characterization of analogous molecules without these inherent and undesirable effects.

Among them, capsaicin (8-methyl-N-vanillyl-6-nonenamide) is a primary pungent and irritating substance present in chilies and red peppers, which are widely used as spices. Capsaicin evokes numerous biological effects and therefore has been the target of extensive investigations since its initial identification in 1919 by E. K. Nelson.

Different pharmacological effects have been attributed to capsaicin since ancient times, but it was not until the past twenty years when extensive research has been done to determine its specific applications.

In this chapter the authors will review the last results obtained in research, considering their application to Veterinary Medicine, not only for its therapeutic role (pain, arthritis, muscle injury, obesity, anthelminthic) but also as a preventive tool for digestive diseases, such as ruminal acidosis or gastric pain. In addition, supplementation of this compound in combination with other plant extracts has been proven to contribute improving animal production in ruminant livestock.

Chapter 3 - Capsaicinoids are a mixture of low molecular weight acid amides and vanillylamine and C_8 to C_{13} fatty acids that belong to the family of vanilloids. The major capsaicinoids naturally occurring in red chili peppers are capsaicin, dihydrocapsaicin and nordihydrocapsaicin, among other chain analogs that also occur in smaller amounts. Capsaicinoids have been reported to influence specific functions such as pain, body temperature regulation, body weight regulation, blood circulation and to exert a protective action to the integrity of the gastric mucosa. Recently, it has also been discovered that vanilloids are able to interfere with the cellular tight junctions (TJ) of several epithelial tissues. This makes these compounds potential candidates to enhance permeability of co-administered drugs across biological barriers like the blood-brain barrier or the intestinal barrier. Several *in vitro* studies have been carried out to investigate the interaction between capsaicin and the TJ of different cell lines. Complementary to permeability experiments, also electrical impedance measurements of cell monolayers and other assays were conducted. Based on these findings, the first attempts have been made to develop drug delivery systems which exploit the permeabilzing properties of the vanilloids. The scope of this review chapter is on an updated overview of the vanilloids which have been documented to interfere with cellular TJ, the proposed underlying mechanisms along with the main investigative methods and potential first applications of a recently discovered action of capsaicinoids.

Chapter 4 - The analgesic property of Capsicum frutescens LINN. SOLANACEAE] extract-CFE (98% Capsaicin by NMR) of South African origin in mice (20-30 g), using 'hot plate' and 'acetic acid' methods was investigated. In the hot plate test, twelve mice in each of 5 groups were administered intraperitoneal (i.p.) CFE or synthetic capsaicin (Fluka Biotechnika-CF) at 0.5-8 mg/kg and at hourly intervals; at 40 C, the mean reaction time (MRT) was estimated. In the 'acetic acid' test method, CFE or CF was administered imp. at various doses (0.5- 8 mg/Kg) at hourly intervals following single treatment using 0.2 mls of 3 % v/v acetic acid-induced writhing responses, or non-treated (n=12). Separate groups of rats (n=12) were pre-treated with 100-mg/kg diclofenac (DIC) or morphine (MPN) 10 mg/kg i.p. to evaluate the peripheral analgesic effects and central analgesia, respectively. Data obtained were pooled and analysed by the student's T-test. The 'hot plate' and 'acetic acid' test methods showed significant ($p<0.0001$) prolongation of the MRT and inhibition of writhing responses, respectively, compared to the control. Similar results with a Pearson correlation of 0.999 were obtained in the CF subgroups. This study shows that capsaicin has superior central analgesia compared to equipotency doses of MPN or DIC and comparable peripheral analgesia to either MPN or DIC.

Chapter 5 - *Background:* The analgesic effect of capsaicin (the active ingredient in *Capsicum frutescens* Linn. *[*Solanacea*e])* had been reported in several studies. Current research is being directed at producing analgesics, anti-inflammatory agents with better side effect profile.

Objectives: To investigate if either the ethyl acetate extract of *Capsicum frutescens* Linn. *[*Solanacea*e]* (CFE) or capsaicin (Fluka Biotechnika-CPF) (in addition to the known analgesic properties) has any anti-inflammatory effect comparable to non-steroidal anti-inflammatory analgesics (NSAIDS).

Methods: The effects of ethyl acetate extract of *Capsicum frutescens* Linn. *[*Solanacea*e]* (CFE) and capsaicin (Fluka Biotechnika-CPF) was examined on rat hind paw. Inflammation was induced in the rat's hind paw by subplantar injections of fresh egg albumin (0.5 ml/kg). Diclofenac (100 mg/kg) was used as the reference anti-inflammatory agent for comparison, while distilled water was used as the placebo. The leucocytes count, corticosterone and C - reactive protein (CRP) levels were measured as biomarkers of inflammation. Data obtained were pooled and analysed using repeated ANOVA, in a general linear model with the CPSS software.

Results: Sub-plantar injections of fresh egg albumin (0.5 ml/kg) produced profound and time-related oedema in the rat hind paw of the 'control' rats. Diclofenac (DIC, 100 mg/kg, i.p.) and reference capsaicin (CPF, 2.5 mg/kg, i.p.) significantly inhibited paw swelling at ($p\leq0.05$–0.001) (CI 95%) compared to distilled water-treated 'controls'. While the corticosterone levels were all very low in 7 rats treated with capsaicin, the leucocytes count was within normal range in 9 rats. However, in 16 specimens randomly assigned for CRP levels, there were very high CRP readings, up to a magnitude of 10 times the normal range.

Conclusion: Capsaicin in both forms (CFE and CPF) produced anti-inflammatory effects that were comparable to diclofenac in the experimental rat model at $p<0.05$. It may be concluded that capsaicin has both analgesic and anti-inflammatory properties.

Chapter 6 - Capsaicin (trans-8-methyl-N-vanillyl-6-nonenamide) is a naturally occurring alkaloid derived from plants of the genus Capsicum. TRPV1 (transient receptor potential vanilloid-1), a capsaicin receptor, is expressed in primary sensory neurons of dorsal root ganglia and vagal nerves. Heat and protons as well as capsaicin activate TRPV1 to induce the influx of cations, particularly Ca^{2+} and Na^{+} ions. Characteristic effects of capsaicin are the induction of a burning sensation after acute administration and the desensitization of sensory neurons after large doses and prolonged administration. Also, activation of capsaicin-sensitive terminals enhances release of neuropeptides, for example substance P and calcitonin gene-related peptide. The use of capsaicin in animal models is important in the study of both processing and modulation of pain signals. Systemic treatment with the capsaicin is known to establish a long-lasting depletion of certain populations of afferent neurons and nerve fibers. Capsaicin induced different morphologic changes when given in different doses and at different ages. Administration of capsaicin modified age-related changes in the number of sensory neurons, containing TRPV1-receptors, calcitonin gene-related peptide, substance P, neurofilament 200.

Chapter 7 - Capsaicin is widely consumed as a spice and food additive, and is also utilized in medical applications. Capsaicin (N-[4-hydroxy-3-methoxyphenyl] methyl-8-methyl-non-6-enamide) and dihydrocapsaicin (N-[4-hydroxy-3-methoxy-phenyl]methyl-8-methyl-non-6-nonanamide) are two members of the family of naturally occurring capsaicinoids, which make up the pungent components of hot pepper. Capsaicin has been obtained in the oleoresin capsicum form through the industrial extraction of the dried ripe fruits of capsicums, and it contains a complex mixture of essential oils, waxes, colored materials, and several capsaicinoids. Capsaicin is one of the carotenoids, such as capsanthin, cryptoxanthin, carotene and zeaxanthin, which contribute to the yellow to red colours of many foods. However, there are several limitations for its use in food systems, such as lipophilicity and the structure of carotenoids, which affect their distribution in the cellular systems *in vivo*, and high melting point, making them crystalline at food storage and body temperatures. One of the promising nanotechnologies is the nanometer-sized emulsion delivery system, which offers the potential to improve the solubility and bioavailability of many bioactive ingredients such as carotenoids, polyunsaturated fatty acids, phytosterols, antioxidants, vitamin, minerals, and other natural compounds. Bioactive ingredients can be encapsulated in nanometer-sized structures, which function as the transport systems. These nanometer-sized structures have sizes ranging from 20–500 nm, and provide the necessary physical properties, such as solubility, stability, and bioavailability. A nanopolymeric system can be applied to the formulation of nanoemulsions through self-assembly emulsification for the development of a bioactive ingredient delivery system. The incorporation of alginate and chitosan to form multi-layer nanoemulsions can be expected to improve the stability of nanoemulsions, and this system holds promise for use in the production of functional foods containing bioactive ingredients. The nanoemulsions of capsaicin which incorporated with alginate (AN) were found to reduced the body weight and adipose tissue mass, whereas capsaicin extract (CE) did not. CE and AN reduced the levels of mRNA of adipogenic genes, including peroxisome proliferator-activated receptor-γ, sterol regulatory element-binding protein-1c, and fatty acid-binding protein in white adipose tissue. The mRNA levels of genes related to β-oxidation or thermogenesis, including PPAR-α, palmitoyltransferase-1α, and uncoupling protein-2, were increased by supplementation with CE and AN relative to the high fat control group. In particular, the expression of PPAR-α, CPT-1α, and UCP2 was well as the activity of AMPK were significantly increased in the AN group compared to those in the OC group.

Chapter 8 - The protective role of capsaicin on oxidative stress produced during acetaminophen induced toxic hepatitis was studied in rats.

Administration of acetaminophen (800 mg/kg body weight intraperitoneally) to rats caused a significant increase in oxidative stress indicated by abnormal changes in lipid peroxidation, enzymic antioxidants (superoxide dismutase, catalase, glutathione peroxidase, glutathione reductase and glutathione-S-transferase) and non-enzymic antioxidants (reduced glutathione, vitamin C, vitamin E and vitamin A).

Treatment with capsaicin (10 mg/kg body weight intraperitoneally) reverted all the above abnormal changes and restored cellular normalcy suggesting the modulatory efficacy of capsaicin in curbing the oxidative stress produced during acetaminophen induced hepatitis.

In: Capsaicin
Editor: Bruce Gilliam
ISBN: 978-1-61470-433-1

Chapter 1

CREATIVE DESTRUCTION IN THE PERIPHERAL NERVOUS SYSTEM: THE ACTION OF CAPSAICIN ON PRIMARY SENSORY NEURONS

Krzysztof Czaja**
Washington State University, Integrative
Physiology and Neuroscience, Pullman WA, US

ABSTRACT

The book chapter titled: "Creative Destruction in the Peripheral Nervous System: The action of capsaicin on primary sensory neurons", includes research on changes in the sensory component (SN) of the peripheral nervous system (PNS), which are the result of the neurotoxic actions of capsaicin. Capsaicin (8-methyl-N-vanillyl-6-nonenamid) is an alkaloid responsible for the spicy, burning taste of chili peppers. In mammals, it creates a burning sensation in the tissue when it comes in contact. Capsaicin works by the vanilloid receptor (TRPV1) and has neurotoxic properties. Intraperitoneal injection of capsaicin has become an appropriate tool to damage the sensory afferent fibers innervating the visceral organs, originating from the nodose ganglion (NG) and the dorsal root ganglia (DRG). Intraperitoneal injection of capsaicin caused the death of about half of the neurons in the studied ganglia.

* Corresponding author: Krzysztof Czaja. E-mail: czajak@vetmed.wsu.edu.

A surprising, and very intriguing, fact is after about two months after the injury, the number of ganglion neurons was restored to the level observed in control animals. The above observation allowed us to test the hypothesis that the capsaicin-induced neuronal death in sensory ganglia activates a cascade of events leading to proliferation of ganglion cells and formation of new neurons, which restore the damaged peripheral sensory pathways. Further studies revealed the presence of nestin, a marker of stem cells in the satellite glia cells (SG), which proves their proliferative potential. SG, under the influence of events induced by capsaicin, enter the cell cycle, synthesize DNA (BrdU incorporation) and show immunoreactivity against Ki-67, an antigen present in the active phases of the cell cycle. Both the central axons from NG neurons forming synaptic connections in the medulla and the peripheral which innervate the upper gastrointestinal tract, are rebuilt after sixty days following the capsaicin administration. We used intraperitoneal administration of cholecystokinin (CCK), which is specifically acting via the vagus nerve to reduce appetite. Our results showed that capsaicin significantly reduced the effect of CCK, which confirmed the damage to the peripheral axons of the vagus nerve. However, three hundred days after injury, the effect of CCK was restored. In addition, and very interesting, is the fact that at this time, the number and immunohistochemical characterization of neurons in the NG returned to that observed in control animals. In summary, our studies clearly show that capsaicin kills a significant population of viscerosensory neurons in PNS but also induces a cascade of events leading to an activation of endogenous multipotent cells in the sensory ganglia. These cells enter the cell cycle and divide and differentiate to form new glia and functional neurons. However, the molecular mechanism of the neurogenic action of capsaicin needs further investigation. Complete understanding of neurogenic capsaicin action and the ability to manipulate the population of endogenous ganglionic progenitors is a fundamental step towards the use of alternative reserve of stem cells to repair a damaged nervous system. This knowledge will help to develop methods to repair damaged CNS, arising as a result of an illness or accident.

INTRODUCTION

The chapter titled: "Creative Destruction in the Peripheral Nervous System: The action of capsaicin on primary sensory neurons", includes research on changes in the sensory component of the PNS, which are the result of the neurotoxic action of capsaicin.

Capsaicin, and several related compounds, are called capsaicinoids and are produced as secondary metabolites by chili peppers. There are five naturally occurring capsaicinoids: capsaicin, dihydrocapsaicin, nordihydrocapsaicin, homocapsaicin, and homodihydrocapsaicin (Reilly et al. 2001). Capsaicin (trans-8-methyl-N-vanillyl-6-nonenamide; Figure 1) is an alkaloid responsible for the spicy, burning taste of chili peppers. Capsaicin is a crystalline, lipophilic, colorless and odorless alkaloid with the molecular formula $C_{18}H_{27}NO_3$. Its molecular weight is 305.40 g/mol, and it is fat-, alcohol- and oil-soluble. It is widely consumed in spicy foods throughout the world, particularly in South East Asia and Latin-America. The capsaicin content of hot red peppers ranges from 0.1 to 1.0% (Korel et al. 2002). In mammals, it creates a burning sensation in the tissue, in which it comes in contact. Capsaicin works via the transient receptor potential vanilloid subtype 1 (TRPV1) (Caterina et al. 1997). TRPV1 is a heat-activated calcium channel that opens between a 37°C and 45°C temperature. When capsaicin acts on TRPV1, it causes the channel to open in a temperature lower than 37°C, so the sensation of heat is created. Capsaicin exerts its major pharmacological effects on the peripheral part of the sensory nervous system (PNS), particularly on the C-type and Aδ-type primary sensory neurons expressing a TRPV1 receptor (Caterina et al. 1997; Hiura et al. 1999). In the PNS a majority of primary sensory neurons express the TRPV1 receptor (Berthoud et al. 1997).

Because of its potential as a toxin for peripheral unmyelinated fibers, capsaicin has been used extensively as a versatile tool in the study of pain mechanisms (Caterina et al. 1997; Derry et al. 2009; Lauria et al. 2006; Rashid et al. 2003) and gut-brain communication (Berthoud et al. 2001; Campos et al. 2012; Carobi 1996; Czaja et al. 2008; Gillespie et al. 2005; Peles et al. 2009; Ritter et al. 1989; Ryu et al. 2010; Zafra et al. 2003).

However, recent studies demonstrated capsaicin-induced dendritic degeneration, in myelinated neurones, from rat dorsal motor nucleus of the vagus (DMV), the source of vagal efferent neurons (Browning et al. 2013).

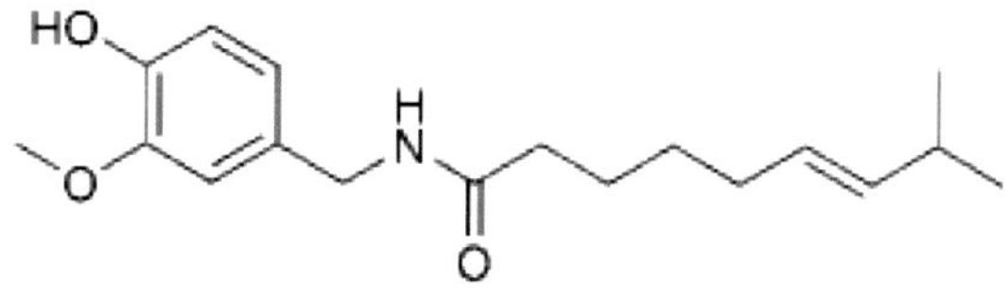

Figure 1. The molecule of capsaicin contains aromatic ring, amide bond and hydrophobic side chain.

Therefore, use of capsaicin as a neurotoxin for afferent C-fibres requires rigorous controls.

The nature and extent of capsaicin's effects on primary afferent neurons depend on toxin dose, route of administration and duration of exposure. Generally capsaicin's effects range from neuronal excitation, sensory blocking (desensitization), degeneration of terminals and/or axons and destruction of nerve cell bodies (Szolcsanyi and Mozsik 1984). It has been used as an injury-inducing tool in both the vagal and the spinal component of the PNS (Figure 2). There are numerous reports of capsaicin-induced primary afferent terminal degeneration, there are only a few reports of the effects of capsaicin treatment on primary afferent perikarya, and most of the studies have involved capsaicin administration in neonates (Carobi 1996; Hiura 2000; Szoke et al. 2002).

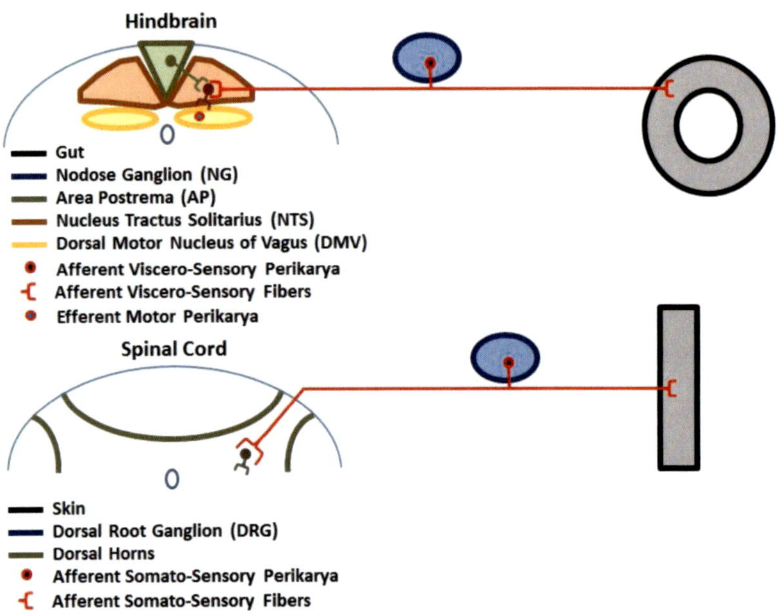

Figure 2. The vagal primary sensory neurons (red, upper) terminate in the Hindbrain (NTS). They provide a sensory innervation to visceral organs (gut for example). The cell bodies of viscero-sensory neurons are located in NG. The spinal primary somatosensory afferents (red, lower) terminate in the Spinal Cord (dorsal horns). The cell bodies of spinal sensory afferent neurons are located in the DRG. The vagal and the spinal primary sensory neurons do not have dendrites. These pseudo unipolar neurons contain an axon that has two branches: peripheral to the target organ and central to the brain or spinal cord.

When capsaicin is applied at the neonatal stage, it acts as neurotoxin, inducing irreversible and specific damage to C-type primary sensory neurons. Neonatal capsaicin treatment of vagal afferents *in vivo* reduced the number of neurons in nodose ganglia by about 70% (Carobi 1996). Several studies on capsaicin action in the PNS have been performed on the somatic component (Figure 2). They revealed that a subpopulation of small-to-medium sized primary sensory neurons, representing about 17% of the total neuronal population in the 4th lumbar dorsal root ganglion, underwent rapid degeneration after the administration of capsaicin (Jancso et al. 1985).

Moreover, they reported a decrease of about 45% in the number of unmyelinated axons in the saphenous nerve. Systemic application of capsaicin in neonatal rats resulted in the death of small primary afferent spinal neurons and the degeneration of unmyelinated fibers in a dose-dependent manner (Hiura et al. 1999; Hiura 2000; Hiura et al. 2002; Nagy et al. 1983). This damage translates to a shift in the size distribution of dorsal root ganglia (DRG) cells toward larger neurons and myelinated fibers later in life (Hiura 2000). It has been previously reported that capsaicin also induces degeneration of peripheral afferents in skin, cornea, urinary bladder and gut (Holzer et al. 1982; Fujita et al. 1987; Hiura et al. 1999). However, the ability of primary sensory neurons damaged by capsaicin to regenerate afferent fibers is still not completely understood. It has been reported that capsaicin treatment results in long-term losses in nociceptive fields and C-fiber conduction (Chung et al. 1990). Conversely, it has been also shown that systemic pretreatment with capsaicin results in increased fiber outgrowth in DRG cultures (Winter et al. 1992). The above studies indicate that capsaicin administration causes axonal degeneration in neonatal and adult rats, but evidence that capsaicin treatment results in destruction of primary afferent neuron cell bodies, especially in adult animals, is equivocal and incomplete. Studies conducted in numerous research centers around the world have shown that afferent sensory fibers supplying the gastrointestinal tract originate from neurons located in the NG (Figure 2) and from sensory neurons of the dorsal root ganglia (DRG) at the thoracic and lumbar level (Gallaher et al. 2010). Our results showed that the majority of sensory neurons in the NG contain ionotropic glutamate receptors (NMDA). However, the structure of the NMDA receptor is not homogeneous (Czaja et al. 2006a; Czaja et al. 2006b). Immunoreactivity of the NMDA NR1 subunit was observed in 92.3% of the population of neurons located in the NG. The NR2B subunit was present in 56.7% of the neurons, NR2C was observed in 49.4% of the population and the NR2D subunit was present only in 13.5% of the total population of neurons in the NG.

These studies showed that the primary sensory neurons located in the NG can be activated by glutamate, which is the main neurotransmitter in the CNS. Lack of homogeneity of the NMDA receptor found in our study also points to the possibility of specialization dependent on the phenotype in the transmission of sensory signals and NMDA phenotype-specific sensitivity to capsaicin action.

Studies of the spinal primary sensory neurons, innervating the upper gastrointestinal tract, showed that the vast majority of afferents originate from thoracic (T8-T13) and lumbar (L1-L2) DRGs (Gallaher et al. 2010). In addition, over 80% of the population expressed an NR1 subunit of an NMDA receptor, which suggests that glutamate is also an important neurotransmitter in the spinal sensory component of the PNS.

Both previously published studies, on the DRG and NG and our results, showed that a large population of primary sensory neurons (~ 60% - ~ 80%) expresses a TRPV1 receptor and is sensitive to capsaicin (Czaja et al. 2008) and contains the sodium ion channel NAV1.8 (Gallaher et al. 2010).

In summary, our studies showed that the vast majority of primary sensory neurons, innervating intraperitoneal organs, is represented by the population of unmyelinated neurons, sensitive to capsaicin. They also showed that glutamate can activate these neurons, becoming an important neurotransmitter in the peripheral sensory system. Therefore, the role of capsaicin in glutamate signaling requires further investigations.

The next step in our studies was to examine long-term damage-related changes in primary sensory neurons following capsaicin treatment. Due to the fact that most primary sensory neurons contain TRPV1 (Berthoud et al. 1997), intraperitoneal injection of capsaicin has become an appropriate tool to damage the sensory afferent fibers innervating visceral organs, originating from the NG and the DRG. We used a neurotoxic dose of capsaicin (125 mg/kg) to cause not only activation but also death of primary sensory neurons. An intraperitoneal injection of capsaicin caused the death of about half the neurons in the NG (Czaja et al. 2008) and DRG (Gallaher et al. 2010) and was observed a month after the administration of the neurotoxin. Capsaicin-induced neuronal death was accompanied by loss of immunoreactivity against TRPV1, NMDA, and Nav1.8 in NG (Gallaher et al. 2011; Czaja et al. 2008) and DRG (Gallaher et al. 2010; Gallaher et al. 2014). In the NG, TRPV1 receptor expression was decreased from 60% to 10% at three days post treatment (Czaja et al. 2008). While in DRG we observed the significant gradual decrease in TRPV1 mRNA levels from day 1 till day 15 after the capsaicine administration (Gallaher et al. 2014).

Our studies also revealed that capsaicin treatment gradually reduced the number of the NR1 subunit of the NMDA receptor-expressing neurons in NG up to 180-days post-injection (Gallaher et al. 2011). We found Nav1.8 to be present in 40% of spinal viscerosensory afferents under normal conditions, consistent with previous reports (Decosterd et al. 2002). Three days after capsaicin treatment, both mRNA and immunoreactivity for Nav1.8 were reduced by approximately half. Because Nav1.8 is preferentially expressed in C- and Ad-type neurons that also express the capsaicin receptor (Amaya et al. 2000; Djouhri et al. 2003), this finding is not unexpected.

And finally, our recently published stereological quantification, of neurons in the right fifth lumbar dorsal root ganglion, showed that the number of neurons in the DRG is significantly reduced 30 days post-capsaicin (Gallaher et al. 2014). We also observed an activation of the apoptosis cascade revealed by an increased expression of cleaved caspase-3. Damaged neurons revealed a nuclear decompartmentalization.

The number of neurons, prelabeled by the retrograde tracer fast blue (FB), injected into the sciatic nerve, seven days prior to administration of capsaicin, showed that prelabeled neurons were destroyed within 30 days of capsaicin administration. Real-time RT-PCR results for the neuronal cytoskeletal protein b-III tubulin, the capsaicin receptor TRPV1, and caspase-3, reflected the loss of capsaicin-sensitive neurons through apoptosis.

In summary, the previously published results and studies from our laboratory show that capsaicin treatment results in significant damage of primary sensory neurons and death of at least 50% of vagal and spinal afferent neurons in NG and DRG (respectively). The damaging force of capsaicin in the sensory division of the PNS is well established, but the cascade of events following the neuronal damage and death needs further investigations.

Our studies aimed at determining the long-term effect of capsaicin, on sensory neurons, brought surprising and very intriguing results. We found that the loss of SN was not permanent. In fact, about two months after the capsaicin-induced neuronal death in both the NG and The DRG, the number of ganglion neurons was restored to the level observed in control animals (Gallaher et al. 2010; Gallaher et al. 2011; Gallaher et al. 2014; Czaja et al. 2008). By 180 days, after systemic capsaicin treatment, the numbers of neurons in NG of capsaicin-treated rats exceeded the numbers observed in control rats (Gallaher et al. 2011). This restoration in the number of vagal sensory neurons in NG led us to hypothesize that a significant population of newly differentiated neurons replaced neurons that died following capsaicin treatment.

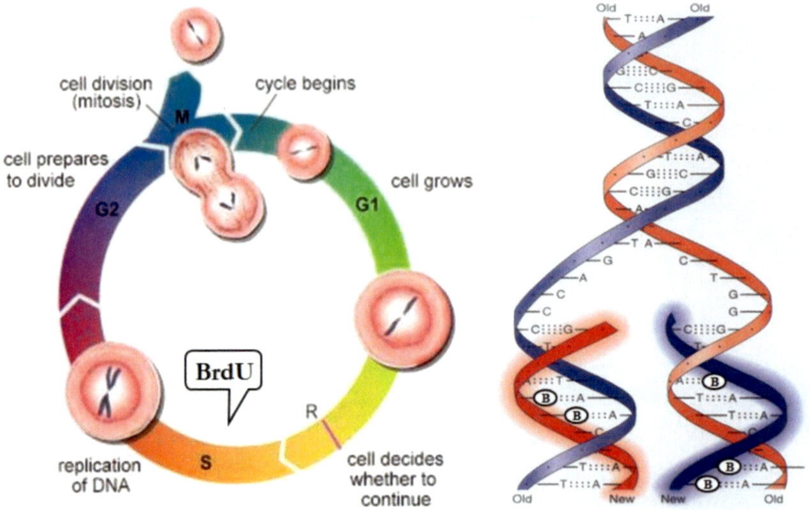

Figure 3. Bromodeoxyuridine (5-bromo-2'-deoxyuridine, BrdU) is a synthetic nucleoside that is an analog of thymidine. BrdU is commonly used in vivo for detection of proliferating cells. BrdU can be incorporated, during the S phase of the cell cycle, into the newly synthesized DNA of replicating cells, replacing thymidine during DNA replication. Antibodies can then be used to detect the incorporated BrdU, labelling cells that were actively replicating their DNA.

Further investigations were designed to determine the phenotype changes in a ganglionic environment following capsaicin treatment. We found that after a period of sixty days the total number of sensory neurons in the ganglia, representing different subpopulations (TRPV1, NMDA and Nav1.8), returned to a level that was observed in control animals (Gallaher et al. 2011; Gallaher et al. 2014; Czaja et al. 2008). The described phenomenon can be interpreted in two ways. First, the damaged neurons can change their morphology and immunoreactivity, and not be recognized by the applied markers and antibodies against neuronal antigens tested. This damage-induced loss of immunoreactivity was previously reported in the CNS (Bywood and Johnson 2000). However, we also found that a significant number of neurons were labeled with markers of cell death, TUNEL and caspase (Czaja et al. 2008).

Moreover, thirty and sixty days after the capsaicin treatment we observed an increased incorporation of 5-bromo-2-deoxyuridin (BrdU), a synthetic thymidine analogue, used as a marker of DNA synthesis (Takahashi et al. 1992; Nowakowski et al. 1989; Hayes and Nowakowski 2002; Caviness, Jr. et al. 2003) (Figure 3). This discovery suggested a second, much more exciting interpretation of the results.

The above observation allowed us to test the hypothesis that the capsaicin-induced death of primary sensory neurons in the ganglia (NG and DRG) activates a cascade of events leading to proliferation of endogenous cells and formation of new neurons, which restore the damaged peripheral sensory pathways. This hypothesis, although very revolutionary, was suggested by earlier *in vitro* studies (Arora et al. 2007).

However, the source and the phenotype of cells that restore neuronal number in sensory ganglia, following capsaicin treatment, was not known. Pluripotent precursor cells might be present in the adult NG or they might differentiate from resident glia following capsaicin treatment. Populations of neurons expressing nestin, a marker related to maturation of neuronal precursors, were previously reported in tri-geminal ganglion of adult rats (Lagares et al. 2007).

Additionally, precursor cells with similar gene expression profiles to those reported in trigeminal ganglion have also been shown in dorsal root ganglia in the adult rat (Li et al. 2007).

Alternatively, it is conceivable that precursors migrate to the ganglion from other locations and then differentiate into neurons. A third possibility is that some neurons of the adult NG retain the potential to divide and generate new neurons after their axons are lost due to capsaicin action. Induced division of neurons derived from the stem cells has been previously reported in vitro (Lin et al. 2004).

In vivo studies conducted in our laboratory clearly indicate that, in addition to regeneration of damaged neurons by capsaicin, which survive and rebuild the damaged axons (Ryu et al. 2010), a significant population of ganglion cells, most likely so-called satellite glia cells (SG), is activated to divide (Figure 4) and differentiate in glia cells or neurons (Gallaher et al. 2011; Gallaher et al. 2014; Czaja et al. 2008).

Further studies revealed the presence of nestin, a marker of stem cells in the SG, which proves their proliferative potential (Gallaher et al. 2011). SG under the influence of events induced by capsaicin enter the cell cycle, synthesize DNA (BrdU incorporation) and show immunoreactivity against Ki-67, antigen present in the active phases of the cell cycle (Gallaher et al. 2011). Moreover, we found that both the central axons from NG neurons forming synaptic connections in the medulla and the peripheral which innervate the upper gastrointestinal tract, are rebuilt after sixty days following the injury (Ryu et al. 2010; Gallaher et al. 2011).

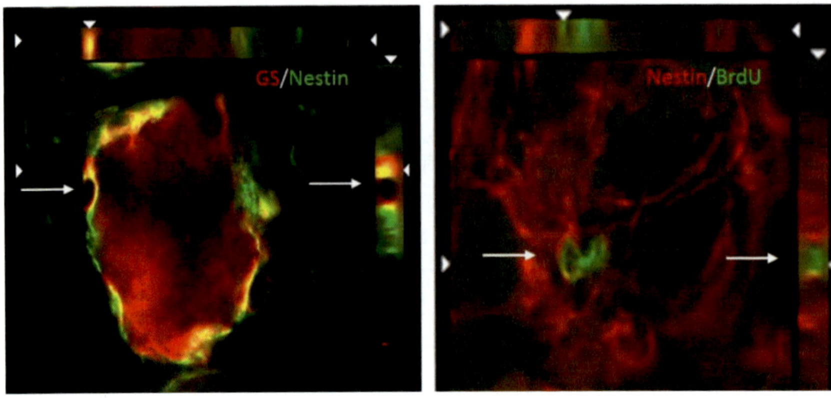

Figure 4. Satellite glia cells (arrows) in the DRG co-express the glial cell marker glutamine synthetase (GS; red, left) and the progenitor cell marker nestin (green, left). Capsaicin treatment induced proliferation of satellite glia reflected by BrdU incorporation (green, right) in nestin-positive cells (red, right).

In the most recent study, we injected a retroviral vector containing the sequence for GFP directly into the fifth lumbar DRG to determine the phenotype of dividing cells in the DRG after capsaicin treatment (Gallaher et al. 2014). This replication-incompetent vector is incorporated into the DNA only if a cell is undergoing mitosis (Lewis and Emerman 1994). However, once the gene has been incorporated, it is stably expressed and passed on to any further progeny. We injected the viral vector at one day or 15 days post-capsaicin to catch the time points from the BrdU experiment that showed the maximal amount of proliferation and when proliferation had returned to baseline, respectively. By this method, we showed a significant increase in GFP-expressing cells one day following systemic administration of capsaicin. The highly correlated increase in proliferation at early time points between resident macrophages and SGs and the potential overlap in the roles of these two cell populations suggest that initial SG proliferation serves to support damaged neurons before apoptotic pathways take over. Although SGs, present in sensory ganglia, express nestin and, following capsaicin treatment, increase proliferation prior to increases in neuronal numbers, the precise role of these cells remains unknown. Future work should be aimed at identifying the specific factors responsible for activating these potential neural progenitors.

The next step in our research was to determine whether restored neural connections are functional. We used intraperitoneal administration of cholecystokinin (CCK), which is specifically acting via the vagus nerve to reduce an appetite (Campos et al. 2012; Campos et al. 2013).

Our results showed that capsaicin significantly reduced the effect of CCK, which confirmed the damage to the peripheral axons of the vagus nerve (Gallaher et al. 2011). However, 300 days after injury, the effect of CCK was restored. In addition, a very interesting fact is that, at this time, the number and immunohistochemical characterization of neurons in the NG returned to that observed in control animals.

Moreover, the quantification of paw withdrawal latency in response to a hot plate showed that heat sensation, normally mediated by TRPV1-expressing neurons, was restored at 60 days postcapsaicin. In summary, these results show that following capsaicin-induced neuronal death the PNS harbors progenitor cells that are capable of proliferating and differentiating into new neurons. Subsequently, these newly formed neurons are capable of integrating into the PNS circuitry and contributing to the restoration of lost functions.

In conclusion, the series of studies reviewed in this chapter clearly show the destructive nature of capsaicin with respect to primary sensory neurons. It is revealed by killing about 50% of neurons in sensory ganglia. However, capsaicin also has a creative nature reflected by inducing a cascade of events leading to an activation of endogenous multipotent cells in the sensory ganglia. These cells enter the cell cycle, divide and differentiate to form new glia and functional neurons, integrating into the PNS circuitry. Although the work reviewed above provides convincing evidence for capsaicin-induced repair and neurogenesis in the adult nervous system, the precise mechanism by which this occurs is not understood. One of the key pathways regulating cell proliferation in the nervous system is WNT (wingless-type MMTV integration site family)/ β-catenin signaling (Wexler et al. 2009).

The potential role of WNT signaling following capsaicin-induced damage has previously been suggested (Czaja K et al. 2011).

The complete understanding of molecular mechanisms of neurogenic capsaicin action and the ability to manipulate the population of endogenous ganglionic progenitors is a fundamental step towards the use of this alternative (to embryionic) reserve of stem cells to repair damaged nervous system. This knowledge will help to develop methods to repair damaged CNS, arising as a result of illness or accident.

References

Amaya, F., Decosterd, I., Samad, T. A., Plumpton, C., Tate, S., Mannion, R. J., Costigan, M., Woolf, C. J. (2000) Diversity of expression of the sensory neuron-specific TTX-resistant voltage-gated sodium ion channels SNS and SNS2. *Mol. Cell. Neurosci.* 15:331-342.

Arora, D. K., Cosgrave, A. S., Howard, M. R., Bubb, V., Quinn, J. P., Thippeswamy, T. (2007) Evidence of postnatal neurogenesis in dorsal root ganglion: role of nitric oxide and neuronal restrictive silencer transcription factor. *J. Mol. Neurosci.* 32:97-107.

Berthoud, H. R., Lynn, P. A., Blackshaw, L. A. (2001) Vagal and spinal mechanosensors in the rat stomach and colon have multiple receptive fields. *Am. J. Physiol. Regul. Integr. Comp. Physiol.* 280:R1371-R1381.

Berthoud, H. R., Patterson, L. M., Willing, A. E., Mueller, K., Neuhuber, W. L. (1997) Capsaicin-resistant vagal afferent fibers in the rat gastrointestinal tract: anatomical identification and functional integrity. *Brain Res.* 746:195-206.

Browning, K. N., Babic, T., Holmes, G. M., Swartz, E. M., Travagli, R. A. (2013) A critical re-evaluation of the specificity of action of perivagal capsaicin. *J. Physiol.*

Bywood, P. T., Johnson, S. M. (2000) Loss of tyrosine hydroxylase immunoreactivity in dendrites is a sensitive index of kainic acid-induced damage in rat substantia nigra neurons in vivo. *Neurosci. Lett.* 280:5-8.

Campos, C. A., Shiina, H., Silvas, M., Page, S., Ritter, R. C. (2013) Vagal afferent NMDA receptors modulate CCK-induced reduction of food intake through synapsin I phosphorylation in adult male rats. *Endocrinology* 154:2613-2625.

Campos, C. A., Wright, J. S., Czaja, K., Ritter, R. C. (2012) CCK-Induced Reduction of Food Intake and Hindbrain MAPK Signaling Are Mediated by NMDA Receptor Activation. *Endocrinology* 153:2633-2646.

Carobi, C. (1996) A quantitative investigation of the effects of neonatal capsaicin treatment on vagal afferent neurons in the rat. *Cell Tissue Res.* 283:305-311.

Caterina, M. J., Schumacher, M. A., Tominaga, M., Rosen, T. A., Levine, J. D., Julius, D. (1997) The capsaicin receptor: a heat-activated ion channel in the pain pathway. *Nature* 389:816-824.

Caviness, V. S., Jr., Goto, T., Tarui, T., Takahashi, T., Bhide, P. G., Nowakowski, R. S. (2003) Cell output, cell cycle duration and neuronal specification: a model of integrated mechanisms of the neocortical proliferative process. *Cereb. Cortex*. 13:592-598.

Chung, K., Klein, C. M., Coggeshall, R. E. (1990) The receptive part of the primary afferent axon is most vulnerable to systemic capsaicin in adult rats. *Brain Res.* 511:222-226.

Czaja, K., Czaja, W. E., Giacobini-Robecchi, M., Geuna, S., Fornaro, M. (2011) Injury-induced DNA replication and neural proliferation in the adult mammalian nervous system. DNA Replication and Related Cellular Processes. *Intech*, pp. 259-282.

Czaja, K., Burns, G. A., Ritter, R. C. (2008) Capsaicin-induced neuronal death and proliferation of the primary sensory neurons located in the nodose ganglia of adult rats. *Neuroscience* 154:621-630.

Czaja, K., Ritter, R. C., Burns, G. A. (2006a) N-methyl-D-aspartate receptor subunit phenotypes of vagal afferent neurons in nodose ganglia of the rat. *J. Comp. Neurol.* 496:877-885.

Czaja, K., Ritter, R. C., Burns, G. A. (2006b) Vagal afferent neurons projecting to the stomach and small intestine exhibit multiple N-methyl-D-aspartate receptor subunit phenotypes. *Brain Res.* 1119:86-93.

Decosterd, I., Ji, R. R., Abdi, S., Tate, S., Woolf, C. J. (2002) The pattern of expression of the voltage-gated sodium channels Na(v)1.8 and Na(v)1.9 does not change in uninjured primary sensory neurons in experimental neuropathic pain models. *Pain* 96:269-277.

Derry, S., Lloyd, R., Moore, R. A., McQuay, H. J. (2009) Topical capsaicin for chronic neuropathic pain in adults. *Cochrane Database Syst. Rev.* CD 007393.

Djouhri, L., Fang, X., Okuse, K., Wood, J. N., Berry, C. M., Lawson, S. N. (2003) The TTX-resistant sodium channel Nav1.8 (SNS/PN3): expression and correlation with membrane properties in rat nociceptive primary afferent neurons. *J. Physiol.* 550:739-752.

Fujita, S., Miyazono, Y., Ohba, N. (1987) Capsaicin-induced corneal changes associated with sensory denervation in neonatal rat. *Jpn. J. Ophthalmol.* 31:412-424.

Gallaher, Z. R., Johnston, S. T., Czaja, K. (2014) Neural proliferation in the dorsal root ganglia of the adult rat following capsaicin-induced neuronal death. *J. Comp. Neurol.*

Gallaher, Z. R., Larios, R. M., Ryu, V., Sprunger, L. K., Czaja, K. (2010) Recovery of viscerosensory innervation from the dorsal root ganglia of the adult rat following capsaicin-induced injury. *J. Comp. Neurol.* 518:3529-3540.

Gallaher, Z. R., Ryu, V., Larios, R. M., Sprunger, L. K., Czaja, K. (2011) Neural proliferation and restoration of neurochemical phenotypes and compromised functions following capsaicin-induced neuronal damage in the nodose ganglion of the adult rat. *Front. Neurosci.* 5:12.

Gillespie, B. R., Burns, G. A., Ritter, R. C. (2005) NMDA Channels Control Meal Size via Central Vagal Afferent Terminals. *Am. J. Physiol. Regul. Integr. Comp. Physiol.*

Hayes, N. L., Nowakowski, R. S. (2002) Dynamics of cell proliferation in the adult dentate gyrus of two inbred strains of mice. *Brain Res. Dev. Brain Res.* 134:77-85.

Hiura, A. (2000) Neuroanatomical effects of capsaicin on the primary afferent neurons. *Arch. Histol. Cytol.* 63:199-215.

Hiura, A., Nakae, Y., Nakagawa, H. (2002) Cell death of primary afferent nerve cells in neonatal mice treated with capsaicin. *Anat. Sci. Int.* 77:47-50.

Hiura, A., Nakagawa, H., Koshigae, Y., Yoshizako, A., Kubo, Y., Ishizuka, H. (1999) Age-related changes in the response to thermal noxious heat and reduction of C-fibers by neonatal treatment with capsaicin. *Somatosens. Mot. Res.* 16:115-121.

Holzer, P., Bucsics, A., Lembeck, F. (1982) Distribution of capsaicin-sensitive nerve fibres containing immunoreactive substance P in cutaneous and visceral tissues of the rat. *Neurosci. Lett.* 31:253-257.

Jancso, G., Kiraly, E., Joo, F., Such, G., Nagy, A. (1985) Selective degeneration by capsaicin of a subpopulation of primary sensory neurons in the adult rat. *Neurosci. Lett.* 59:209-214.

Korel, F., Bagdatlioglu, N., Balaban, M. O., Hisil, Y. (2002) Ground red peppers: capsaicinoids content, Scoville scores, and discrimination by an electronic nose. *J. Agric. Food Chem.* 50:3257-3261.

Lagares, A., Li, H. Y., Zhou, X. F., Avendano, C. (2007) Primary sensory neuron addition in the adult rat trigeminal ganglion: evidence for neural crest glio-neuronal precursor maturation. *J. Neurosci.* 27:7939-7953.

Lauria, G., Morbin, M., Lombardi, R., Capobianco, R., Camozzi, F., Pareyson, D., Manconi, M., Geppetti, P. (2006) Expression of capsaicin receptor immunoreactivity in human peripheral nervous system and in painful neuropathies. *J. Peripher. Nerv. Syst.* 11:262-271.

Lewis, P. F., Emerman, M. (1994) Passage through mitosis is required for oncoretroviruses but not for the human immunodeficiency virus. *J. Virol.* 68: 510-516.

Li, H. Y., Say, E. H., Zhou, X. F. (2007) Isolation and characterization of neural crest progenitors from adult dorsal root ganglia. *Stem Cells* 25: 2053-2065.

Lin, Q. X., Que, H. P., Lu, S. H., Liu, S. J. (2004) Induced-division of neurons derived from neural stem cells. *Acta Physiol. Sinica* 56:130-136.

Nagy, J. I., Iversen, L. L., Goedert, M., Chapman, D., Hunt, S. P. (1983) Dose-dependent effects of capsaicin on primary sensory neurons in the neonatal rat. *J. Neurosci.* 3:399-406.

Nowakowski, R. S., Lewin, S. B., Miller, M. W. (1989) Bromodeoxyuridine immunohistochemical determination of the lengths of the cell cycle and the DNA-synthetic phase for an anatomically defined population. *J. Neurocytol.* 18:311-318.

Peles, S., Medda, B. K., Zhang, Z., Banerjee, B., Lehmann, A., Shaker, R., Sengupta, J. N. (2009) Differential effects of transient receptor vanilloid one (TRPV1) antagonists in acid-induced excitation of esophageal vagal afferent fibers of rats. *Neuroscience* 161:515-525.

Rashid, M. H., Inoue, M., Kondo, S., Kawashima, T., Bakoshi, S., Ueda, H. (2003) Novel expression of vanilloid receptor 1 on capsaicin-insensitive fibers accounts for the analgesic effect of capsaicin cream in neuropathic pain. *J. Pharmacol. Exp. Ther.* 304:940-948.

Reilly, C. A., Crouch, D. J., Yost, G. S. (2001) Quantitative analysis of capsaicinoids in fresh peppers, oleoresin capsicum and pepper spray products. *J. Forensic Sci.* 46:502-509.

Ritter, R. C., Ritter, S., Ewart, W. R., Wingate, D. L. (1989) Capsaicin attenuates hindbrain neuron responses to circulating cholecystokinin. *Am. J. Physiol.* 257:R1162-R1168.

Ryu, V., Gallaher, Z., Czaja, K. (2010) Plasticity of nodose ganglion neurons after capsaicin- and vagotomy-induced nerve damage in adult rats. *Neuroscience* 167:1227-1238.

Szoke, E., Seress, L., Szolcsanyi, J. (2002) Neonatal capsaicin treatment results in prolonged mitochondrial damage and delayed cell death of B cells in the rat trigeminal ganglia. *Neuroscience* 113:925-937.

Szolcsanyi, J., Mozsik, G. (1984) Effects of capsaicin on the development of gastric mucosal damage by different necrotizing agents and of gastric cytoprotection by PGI2 atropine and cimetidine on rats. *Acta Physiol. Hung.* 64:287-291.

Takahashi, T., Nowakowski, R. S., Caviness, V. S., Jr. (1992) BUdR as an S-phase marker for quantitative studies of cytokinetic behaviour in the murine cerebral ventricular zone. *J. Neurocytol.* 21:185-197.

Wexler, E. M., Paucer, A., Kornblum, H. I., Palmer, T. D., Geschwind, D. H. (2009) Endogenous Wnt signaling maintains neural progenitor cell potency. *Stem Cells* 27:1130-1141.

Winter, J., Evison, C. J., O'Brien, C., Benowitz, L., Lindsay, R. M., Mulderry, P., Woolf, C. (1992) Neurotoxic damage evokes regenerative responses from adult rat sensory neurones. *Neurosci. Lett.* 146:48-52.

Zafra, M. A., Molina, F., Puerto, A. (2003) Effects of perivagal administration of capsaicin on post-surgical food intake. *Auton. Neurosci.* 107:37-44.

In: Capsaicin
Editor: Bruce Gilliam

ISBN: 978-1-61470-433-1

Chapter 2

NUTRITIONAL AND THERAPEUTIC USE OF CAPSAICIN IN VETERINARY MEDICINE

***Cristina Castillo*[*,3], *Ángel Abuelo*[1], *Jose L. Benedito*[3] and *Joaquín Hernández*[2]**

[1]Resident of the European College of Bovine Health Management (ECBHM)
[2]Diplomate of the European College of Bovine Health Management (ECBHM)
[3]Department of Animal Pathology, Faculty of Veterinary Science, University of Santiago de Compostela, Spain

ABSTRACT

Capsaicinoids are important for the food and pharmaceutical industries. For this reason, a number of researchers are engaged in improving their production by manipulating chili plant cultivation conditions, chemical and enzymatic synthesis processes or alternative methods such as cell or tissue culture. Hitherto, research has shown that capsaicinoids, and capsaicin in particular, have a wide variety of biological and physiological activities, providing them with different functions such as antioxidant, anticarcinogenic, anti-inflammatory, promoter of energy metabolism and suppressor of fat accumulation. However, the potential applications of these molecules are limited by the

[*] Corresponding author: cristina.castillo@usc.es.

irritation caused by their pungency. This has driven the search and characterization of analogous molecules without these inherent and undesirable effects (Reyes-Escogido et al., 2011).

Among them, capsaicin (8-methyl-N-vanillyl-6-nonenamide) is a primary pungent and irritating substance present in chilies and red peppers, which are widely used as spices. Capsaicin evokes numerous biological effects and therefore has been the target of extensive investigations since its initial identification in 1919 by E. K. Nelson.

Different pharmacological effects have been attributed to capsaicin since ancient times, but it was not until the past twenty years when extensive research has been done to determine its specific applications.

In this chapter we will review the last results obtained in research, considering their application to Veterinary Medicine, not only for its therapeutic role (pain, arthritis, muscle injury, obesity, anthelminthic) but also as a preventive tool for digestive diseases, such as ruminal acidosis or gastric pain. In addition, supplementation of this compound in combination with other plant extracts has been proven to contribute improving animal production in ruminant livestock.

INTRODUCTION

According to the *United States Pharmacopeia*, capsicum is defined as the dried ripe fruit of *Capsicum frutescens* Linné (known in commerce as African Chilies), or of *Capsicum annuum* Linné var. *connoides* Irish (known in commerce as Tabasco Pepper), or *Capsicum annuum* var. *longum* Sendt (known in commerce as Louisiana Long Pepper, or a hybrid between the Honka variety of Japanese Capsicum and the Old Louisiana Sport Capsicum, known in commerce as Louisiana Sport Pepper [Fam. Solanaceae]) (see American College of Toxicology, 2007).

The compound was first extracted (albeit in impure form) in 1816 by Christian Friedrich Bucholz (1770–1818). He named it "capsicin", after the genus *Capsicum* from which it was extracted. John Clough Thresh (1850–1932), who had isolated capsaicin in almost pure form, gave it the name "capsaicin" in 1876. But it was Karl Micko who first isolated capsaicin in pure form in 1898. Capsaicin's empirical formula (chemical composition) was first determined by E. K. Nelson in 1919 (Figure 1). This molecule was first synthesized in 1930 by E. Spath and S. F. Darling. In 1961, similar substances were isolated from chili peppers by the Japanese chemists S. Kosuge and Y. Inagaki, who named them capsaicinoids (Reyes-Escogido et al., 2011; Chhabra et al., 2012).

Capsaicin (a phenylpropanoid compound) is the pungent principle of red pepper (*Capsicum annuum* and *Capsicum frutescens*), which is chemically an alkaloid: 8-methyl-N-vanillyl-6-noneamide ($C_{18}H_{27}NO_3$) with a molecular mass of 305.41 g/mol (Börzsei, 2012; Chhabra et al., 2012).

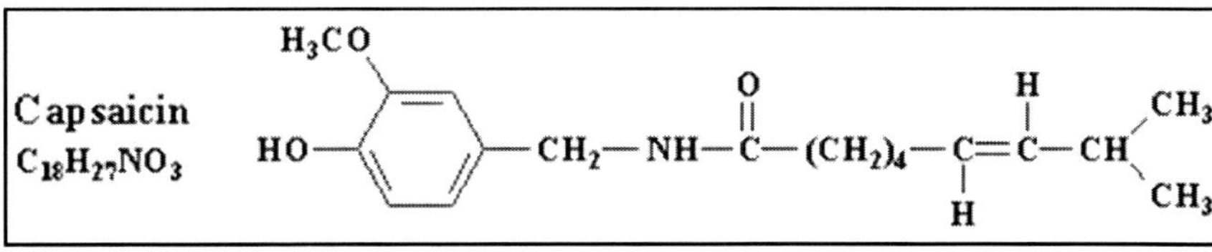

Figure 1. The chemical structure of capsaicin (trans-8-methyl-N-vanillyl-6-noneamide.

Capsaicin displays *cis/trans* isomerism because the double bond prevents internal rotation. Capsaicin is always found as the *trans* isomer because in the *cis* form, the -CH(CH_3)$_2$ and the longer chain on the other side of the double bond will be close together, causing them to repel each other slightly; this steric hindrance does not exist in the *trans* isomer. This additional strain imposed causes the *cis* isomer to be a less stable arrangement than the *trans* isomer (Reyes-Escogido et al., 2011 see Figure 2)

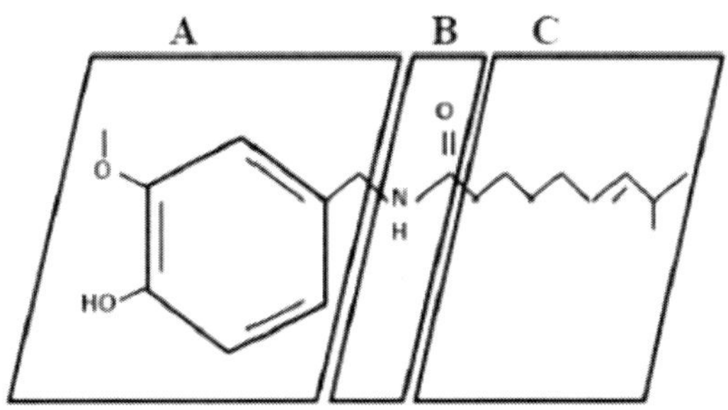

Figure 2. Regions of the molecule of capsaicin. A: aromatic ring; B: amide bond and C: hydrophobic side chain (from Reyes-Escogido et al., 2011).

The pungent compounds of the *Capsicum* fruit are called capsaicinoids (capsaicin and its analogs), and five naturally occurring capsaicinoids have been reported: capsaicin, dihydrocapsaicin, nordihydrocapsaicin, homocapsaicin, and homodihydrocapsaicin (Figure 3).

These capsaicinoids are the major components of most *Capsicum* species, constituting approximately 95% or more of the total capsaicinoid content (Krajewska and Powers 1987). Using mass spectroscopy, the following components of capsaicinoids in *Capsicum annuum* were determined:

Capsaicin (69%), dihydrocapsaicin (22%), nordihydrocapsaicin (7%), homocapsaicin (1%), and homodihydrocapsaicin (1%) (Bennett and Kirby 1968).

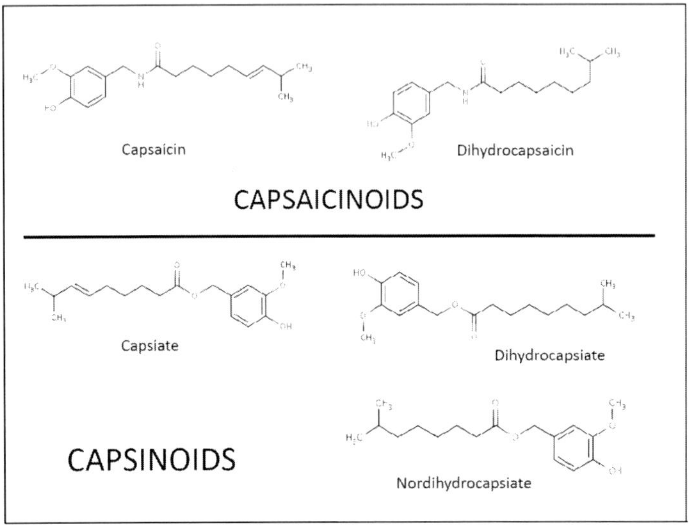

Figure 3. The structures of the capsaicinoids and capsinoids (Barrington, 2014).

A group of structurally related compounds in chillies, the capsinoids, have also been recently discovered. The capsinoids capsiate, dihydrocapsiate and nordihydrocapsiate lack the pungency of the capsaicinoids, and hence the pepper they were first extracted from is called sweet-19 (Reyes-Escogido et al., 2011; Barrington, 2014).

Capsaicinoids are important in the food and pharmaceutical industries. For this reason, a number of researchers are engaged in improving their production manipulating chili plant cultivation conditions, chemical and enzymatic synthesis processes or alternative methods such as cell or tissue culture. Hitherto, research has shown that capsaicinoids, and capsaicin in particular, have a wide variety of biological and physiological activities, providing them with different functions such as antioxidant, anticarcinogenic, anti-inflammatory, promoter of energy metabolism and suppressor of fat accumulation. However, the potential applications of these molecules are limited by the irritation caused by their pungency. This has driven the search and characterization of analogous molecules without these inherent and undesirable effects (Reyes-Escogido et al., 2011).

> When the effect of a specific active component is needed, it is more efficient to use a purified molecule alone than a dried plant or unpurified extract. But we must be aware of the fact that the potency of an unpurified extract often exceeds the potency of a purified one because of synergistic effect among the molecules in it (Frankic et al., 2009).

Different pharmacological effects have been attributed to capsaicin since ancient times, but it was not until the past twenty years when extensive research has been done to determine its specific applications. In this chapter we will review the last results obtained in research, considering their application to Veterinary Medicine, not only for its therapeutic role (pain, obesity, etc.) but also as a preventive tool for digestive diseases, such as ruminal acidosis, or even as growth promoter in livestock.

PHARMACOKINETICS: MAIN ASPECTS

An important factor when considering the exploitation of plant secondary metabolites for the manipulation of peripheral metabolism is bioavailability. Bioavailability depends on the plant secondary metabolites entering the systemic circulation intact after consumption following passage through the gut, across the gut wall and first passage through the liver. The gut, gut wall and liver can extensively metabolize plant secondary metabolites before they reach the systemic circulation (Greathead, 2003).

In an extensive study of capsaicin distribution in tissues, elimination and other active principles in animals with oral administration (Reyes-Escogido et al., 2011), nearly 94% of orally administered capsaicin was absorbed and maximum concentration in the blood was reached 1 h after administration. In addition, a maximum distribution of 24.4% of administered capsaicin in blood, liver, kidney and intestine was seen in 1h and then diminished notably until being undetected after 4 days. Capsaicin metabolism is apparently similar in human, rat and dog microsome. Three major metabolites have been identified for capsaicin: 16-hydroxycapsaicin, 17-hydroxycapsaicin and 16,17-dihydrocapsaicin. Capsaicin is mainly eliminated by the kidneys with a small untransformed proportion excreted in the feces and urine.

> Previous publications suggest that capsaicinoids are metabolized to a great extent by hepatic enzymes (Reilly and Yost, 2006). Early studies demonstrated that capsaicin is converted to metabolites via hydroxylation of the vanillyl ring moiety; leading to the hypothesis that capsaicin is activated

> by the liver mixed-function oxidase system to an electrophilic intermediate that is capable of covalently binding to hepatic proteins. Additionally, the alkyl side chain of capsaicin was also considered to be susceptible to enzymatic oxidation, giving rise to a hydroxylated metabolite at the terminal carbon of the side chain when incubated with NADPH and a liver S9 fraction. One-electron oxidation of capsaicin also plays a role in capsaicin metabolism. Liver cytochrome P450 2E1 (CYP2E1) activity has been shown to be responsible for conversion of capsaicin to the reactive phenoxy radical, which in turn can dimerize or bind to CYP2E1, thereby inactivating the enzyme (Surh and Lee, 1995). This mechanism may be responsible for the inhibition of further activation and the reported chemoprotective activity of capsaicin against some chemical carcinogens and mutagens (Surh and Lee, 1996).

The studies carried out by Chadra et al. (2008) have demonstrated that the metabolism of capsaicin is similar in human, rat, and dog microsomes and S9 fractions. Their *in vitro* results suggest that capsaicin was converted in microsomes and S9 fractions to at least five primary metabolites (Figure 4).

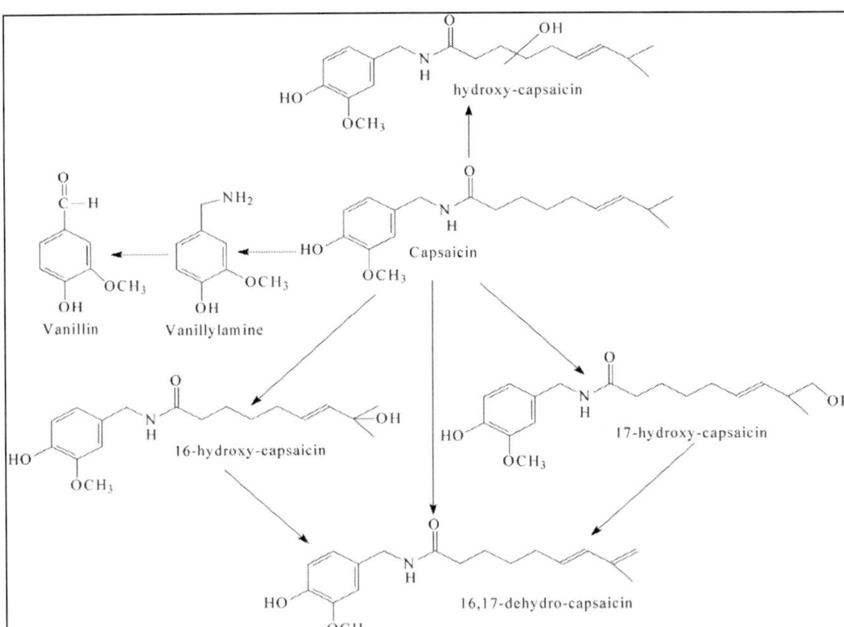

Figure 4. Biotransformation pathway for capsaicin in the liver proposed by Chanda et al. (2008).

Side chain-hydroxylated metabolites were detected including two major metabolites, 16-hydroxycapsaicin and 17-hydroxycapsaicin. Oxidation of capsaicin or loss of water from the hydroxylated metabolites generated 16,17-dehydrocapsaicin. Hydrolysis of the amide bond of capsaicin produced vanillylamine. Further oxidation of vanillylamine generated a secondary metabolite, vanillin. Proposed generation of vanillylamine from capsaicin and subsequent metabolism to vanillin is similar to the metabolic pathway of dihydrocapsaicin suggested by Kawada and Iwai (1985). Additional metabolites detected in dog microsomes and S9 fraction included hydroxycapsaicin. The extent of metabolism was similar in dog microsomes and S9 fractions but was slower than that observed in human and rat.

Capsaicin and Nociception

The use of capsaicin as an analgesic is not new. Hot pepper pods were used by Native Americans to relieve toothaches. In 1850, the Dublin Free Press recommended the use of hot pepper extract as a "remedy for chilblains." Capsaicin has been used more recently in humans for the treatment of cluster headaches, post-mastectomy pain, diabetic neuropathy, post-herpetic neuropathy, and other painful conditions resistant to conventional analgesics (Fusco, 1997; Szallasi and Blumberg, 1999).

Capsaicin, as a member of vanilloid family, binds selectively to vanilloid receptor subtype 1 (VR1).

> Capsaicin is one of several compounds classified as vanilloids because of their characteristic binding of vanilloid receptors in A and C nerve fibers (Seino et al., 2003).

The gene of the capsaicin receptor was identified and cloned in 1997, and the name Vanilloid 1 Receptor (VR1) was given. Later the international nomenclature based on the structure of receptors altered this denomination, and this ligand-gated cation channel was classified as a member of the Transient Receptor Potential (TRP) family, it got the new name of Transient Receptor Potential Vanilloid 1 (TRPV1, Börzsei, 2012). TRPV1 is a non-selective cation channel with polymodal sensor function, which can be activated by a variety of exogenous irritants, as well as endogenous physical or chemical stimuli (Figure 5).

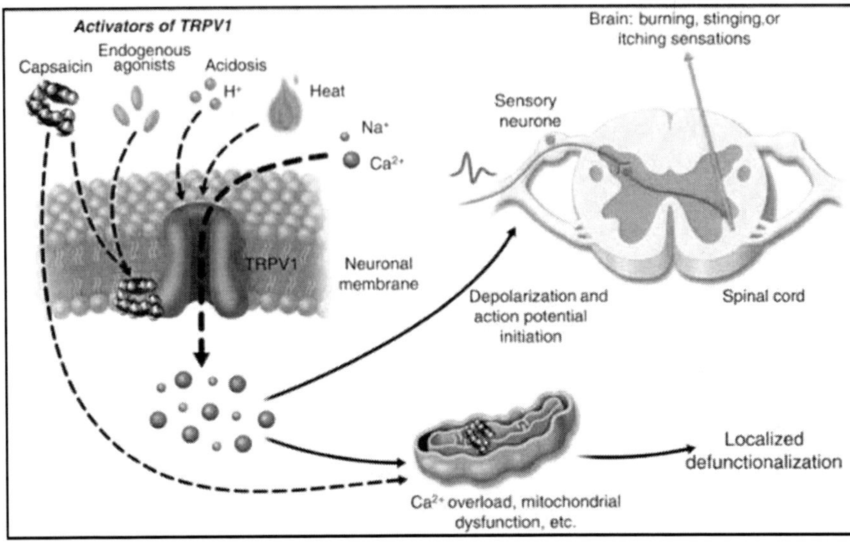

Figure 5. Activation of TRPV1 by capsaicin results in sensory neuronal depolarization (from Anand and Bley, 2011).

> Besides capsaicin, there are other several plant derived substances with vanilloid structure, which can cause receptor stimulation: resiniferatoxin found in Moroccan euphorbia (*Euphorbia resinifera*), piperin in the black pepper (*Piper nigrum*), zigeron extracted from ginger (*Zingiber officinale*) or eugenol of clove (*Syzygium aromaticum*).

By binding to VR1 receptor, the capsaicin molecule produces the same sensation that is produced by excessive heat or abrasive damage; this explains the burning pain caused by capsaicin. Capsaicin also triggers release of the neuropeptide P from the C-type sensory nerve fibers (unmyelinated). Neurons, not having TRPV1 receptors, remain unaffected.

> While sensory neuropeptides are released from the nerve terminals into the innervated area, some peptides like somatostatin and pituitary adenylate cyclase activating polypeptide, are even able to get into the systemic circulation (Börzsei, 2012).

Continuous or repeated activation of the nerve endings results in the loss of their function (desensitization): they cannot be activated by chemical stimuli, but their responsiveness to some physical stimuli remains unchanged. TRPV1 is present in a large proportion of dorsal root and trigeminal ganglia, particularly in small and medium diameter sensory neurons; that is the thinly myelinated A-δ and the above mentioned unmyelinated C-fibres (Börzsei, 2012; Chhabra et al., 2012).

Curiously, the application of capsaicin to human skin also induces neurogenic inflammation characterized not only by transient burning pain but also localized erythema and flare caused by neurogenic vasodilation. Capsaicin-induced neurogenic inflammation is associated with the development of primary thermal and mechanical hyperalgesia at the site of challenge and secondary hyperalgesia in the surrounding area (Di Giminiani et al., 2014).

> This effect is more evident when capsaicin is administered intradermically, causing tissue disruption and leading to sustained changes in sensitivity following initial intense pain (Anderson et al., 2002). On the contrary, capsaicin applied topically induces moderate pain in humans, mostly represented by pruritus (Petersen et al., 2010).

Based on the highly selective agonistic property of capsaicin towards TRPV1 receptors, drug products containing pure synthetic trans-capsaicin are under evaluation as topical and injectable analgesic therapies (Bley 2004).

External Analgesic

On the basis of the ability of capsaicin to activate TRPV1 expressed in nociceptive sensory neurons, various topical and injectable formulations are being developed as potential treatment of various pain syndromes (Anand and Bley, 2011).

Capsaicin is known to be effectively absorbed topically from the skin (Reyes-Escogido et al., 2011).

> In a study of 12 subjects topically administered a 3% capsaicin solution in three different vehicles (70% isopropyl alcohol; mineral oil; and propylene glycol in 20% alcohol), capsaicin was shown to be rapidly absorbed and to quickly reach maximum concentration when capsaicin is applied topically. The half-life of capsaicin was approximately 24 h.

> The implication for topical capsaicin-containing analgesic products is that capsaicin can reach its peripheral target site intact and can reside at the site of action unchanged for a longer period of time. Additionally, a lack of oxidative metabolism in the skin would indicate that there is a potential lack of formation of covalently bound toxic metabolites (Chanda et al., 2008).

Capsaicin (0.025-0.075%) is used in topical ointments to relieve peripheral neuropathies. Capsaicin is the key ingredient of a promising drug under development for postsurgical and osteoarthritic pain. The analgesic effect induced by capsaicin is enhanced in inflammatory states. Capsaicin induces its analgesic effect following an initial excitatory response. It has been demonstrated that the vanilloid system plays an important role in inflammatory hyperalgesia. Capsaicin is currently available as topical ointments, as well as a high-dose dermal patch, which is to be applied for 30-60 minutes, to relieve the pain of peripheral neuropathy. It may be used as a cream for the temporary relief of minor aches and pains of muscles and joints associated with arthritis; often in compounds combined with other rubefacients. The treatment typically involves the application of a topical anesthetic until the area is numb (Chhabra et al., 2012).

Fang et al. (1995) studied percutaneous absorption of hydrophilic and commercial ointment bases containing 0.075% capsaicin using various skin (full-thickness) types.

> The following *in vitro* permeability coefficients (cm/h) were reported in this study for capsaicin (from hydrophilic ointment base): 0.49 ± 0.09 (human skin), 2.91 ± 0.47 (rat skin), 3.92 ± 0.49 (mouse skin), 2.55 ± 0.55 (rabbit skin), and 0.71 ± 0.08 (pig skin). Permeability coefficients (cm/h) for capsaicin (from commercial ointment base) were as follows: 0.88 ± 0.13 (human skin), 3.17 ± 0.13 (rat skin), 8.43 ± 0.51 (mouse skin), 2.96 ± 1.01 (rabbit skin), and 1.27 ± 0.20 (pig skin).

Degim et al. (1999) evaluated the *in vitro* skin penetration of naproxen (nonsteroidal anti-inflammatory agent) in the presence of capsaicin (in ethanol) using full-thickness, human skin (from females, with the underlying fatty tissue removed) that was obtained after abdominal surgery. *Ex vivo* experiments using isolated perfused rabbit ear skin were also performed. Naproxen fluxes were averaged over a period of 60 (human skin) and 6 (rabbit skin) hours. Capsaicin enhanced the skin permeation of naproxen through human skin by approximately twofold. For rabbit skin, an enhancement ratio of 3.8 was reported for naproxen in the presence of capsaicin. No gross changes in structural features of the *stratum corneum* were registered

In veterinary medicine, perineural capsaicin has been used clinically in some equine hospitals as an adjunctive method for relieving pain caused by laminitis. The role of substance P in equine arthritis and foot pain has been extensively studied (Seino et al., 2003). Nerve fibers containing substance P and calcitonin gene–related peptide as well as other neurotransmitting fibers have been shown in the equine navicular bone, proximal sesamoid bones, and various joints, including the middle carpal and the metacarpophalangeal joints (Nixon and Cummings, 1994). Substance P, calcitonin gene–related peptide and vasoactive intestinal peptide have been also found in single nerve fibers that accompany arteries and arteriovenous anastomoses along the length of the dermal laminae of the equine foot (Molyneux et al., 1994). Topical application of capsaicin ointment over the palmar digital nerves reduced pain associated with spontaneous or induced equine laminitis as was demonstrated by Seino et al. (2003).

> Nevertheless, the study showed that the degree of analgesia was not consistent among all horses. This variability might also be due, in part, to a dose-dependent phenomenon. The study also suggests that as substance P can replenish itself after initial depletion, repeated applications of capsaicin may be more effective than a single application.

However, as occurs in humans, vanilloid binding of capsaicin by sensory neurons in horses results in a large influx of calcium ions and then the release of neurotransmitters and nociceptive peptides, manifesting hyperalgesia and pain after application of capsaicin over the digital nerves. Several studies point out that the significant effects of capsaicin in altering lameness should raise concerns over the potential abuse of capsaicin to mask lameness in competition or during pre-purchase examinations. It may be also possible to use capsaicin to create an unfair advantage in competition, especially under nowadays' rigorous zero-tolerance rules.

> Seino et al. (2003) alerts that capsaicin provides sufficient relief from foot pain. Therefore, laminitic horses with capsaicin-induced analgesia might be encouraged to walk, which could result in further mechanical damage to their laminae.

In dentistry, there are studies that have demonstrated that capsaicin increases pulpal blood flow and gives anesthesia to teeth with inflamed pulps, which do not get numb with common anesthetics (Talati and Akbari, 2003). In the study performed by Liu et al. (1992) on cats, they demonstrated that the local use of capsaicin increased the circulation of tooth pulp and that repeated exposure to capsaicin decreased the circulation. However, Madani et al. (2012) observed in cats that capsaicin can decrease the severity of dental inflammation, but it is ineffective in dentinal barrier formation.

In rats, Yoshimura et al. (2000) used a 1% capsaicin cream in their study. At first, it created neurogenic inflammation but then blocked the transmission of pain. Hargreaves et al. (2003) stated that adrenergic agonist due to its direct effect on susceptible pain receptors to capsaicin, relieved the pain and changed the blood circulation through the vascular obstruction and interrupted the transpiration of peripheral neuropeptides.

Antipruriginous

Atopic dermatitis is one of the most common allergies in dogs. Estimated prevalence ranges from 15 to 30% of the canine population. In humans, capsaicin has been used successfully for a variety of pruritic skin conditions such as contact allergy, chronic prurigo, pruritic psoriasis, hydroxyethyl starch-induced pruritus, haemodialysis related pruritus, uraemic pruritus, cold and heat urticarial and aquagenic pruritus (Kennis, 1982; Lundblad et al., 1987; Greaves, 1996). It is hypothesized that capsaicin works by depleting the sensory nerve endings of substance P, implicated in the pathogenesis of atopic dermatitis.

> Substance P acts as a primer of canine neutrophils and may act as a mediator of neurogenic inflammation (Thomsen, 1991).

Nevertheless, capsaicin may be less effective at decreasing pruritus caused by atopic dermatitis than for other pruritic skin diseases (Weisshaar et al., 1998).

> In the report of Weisshaar et al. (1998) researchers found that capsaicin effectively suppressed histamine-induced itching in healthy skin but had less effect in atopic dermatitis.

In the study performed by Marsella et al. (2002) in dogs with atopic dermatitis, authors did not obtained conclusive results about the role of capsaicin in the improvement of pruritus (Table 1). They concluded that it may be due to a mechanism not related to substance P depletion. Indeed, it also remains unknown whether dogs have the capsaicin receptor. In this sense, topical capsaicin continues being evaluated as an adjunctive antipruritic agent in dogs with atopic dermatitis.

But the effectiveness of capsaicin in pruritic conditions is still under investigation, studying other mechanism besides substance P.

Table 1. A: Substance P concentrations in the dog skin before and after treatment for atopic dermatitis. Results are reported as mean ± SEM on a per g tissue weight basis. B: Substance P concentration in lesional and nonlesional skin expressed as mean (pg/g, from Marsella et al., 2002)

A. Substance P (pg/g)		Week 0	Week 6	P values (Week 0 vs. Week 6)
Capsaicin (CAP)		1088.24 ±83.22	1021.12 ±83.22	0.5743
Placebo (PLCB)		987.89 ±83.22	899.51 ±83.22	0.4607
P-values (CAP vs. PLCB)		0.4031	0.3127	
B. Substance P (pg/g)		Week 0	Week 6	P values (Week 0 vs. Week 6)
Capsaicin (CAP)	Lesional skin	955.58	1054.85	0.5570
	Nonlesional skin	1220.91	987.39	0.1746
Placebo (PLCB)	Lesional skin	765.35	850.42	0.6144
	Nonlesional skin	1210.43	948.6	0.1300

Whereas in earlier studies, depletion of substance P was considered the main mechanism of action of capsaicin, it is now accepted that several other mechanisms are involved in the desensitization, including receptor inactivation, blockage of voltage-activated calcium channels, intracellular accumulation of ions leading to osmotic changes, activation of proteolytic enzyme processes, modulation of the production of inflammatory mediators by inhibition of nuclear factor κB activation, and degeneration and reversible loss of nerve fibers (Marsella et al., 2002).

CAPSAICIN AND OBESITY IN PETS

Obesity is defined as an accumulation of excessive amounts of adipose tissue in the body, and is the most common nutritional disorder in companion animals. Studies from various parts of the world have estimated the prevalence of obesity in the dog population to be between 22 and 40%. In another large study based on Australia, 33.5% of dogs were classed as overweight, whereas just 7.6% were judged to be obese. The prevalence of feline obesity was similar (German, 2006).

In humans, obesity is important because it increases mortality risk and can predispose to a variety of diseases. Obese humans, on average, do not live as long, and are more likely to suffer from diseases such as type II diabetes mellitus, hypertension, coronary heart disease, certain cancers (e.g., breast, ovarian, prostate), osteoarthritis, respiratory disease, and reproductive disorders. Similarly, obesity has detrimental effects on the health and longevity of dogs and cats, although data are more limited. Problems to which obese companion animals may be predisposed include orthopedic disease, diabetes mellitus, abnormalities in circulating lipid profiles, cardiorespiratory disease, urinary disorders, reproductive disorders, neoplasia (mammary tumors, transitional cell carcinoma), dermatological diseases, and anesthetic complications. Human obesity is associated with an increased risk of type II diabetes mellitus, cancer, cardiac disease, hypertension, and decreased longevity (Kolonin et al., 2004).

Nearly 20 years ago, another major pharmacological effect of capsaicin was identified when it was found to reduce adipose tissue in rodents by enhancing energy and lipid metabolism, possibly by increasing catecholamine secretion from the adrenal medulla in response to activation of the sympathetic nervous system. These findings have helped to support further research into capsaicin's possible therapeutic effects and mechanisms in obesity treatment (Reyes-Escogido et al., 2011)

Capsaicinoids and capsinoids are of nutritional interest because they show thermogenic properties. Therefore extracts of chilli peppers may be useful adjuncts for weight loss treatments. All the capsaicinoids appear to show some thermogenic properties, but the capsinoids are of particular interest because while being thermogenic they lack the ferocious taste of the capsaicinoids (Barrington, 2014)

Capsaicinoids and capsinoids have been investigated with respect their fat loss effect and found to be effective for long term weight loss. Capsaicinoids and capsinoids can increase thermogenesis and fat oxidation, increasing total energy expenditure (Whiting et al., 2012). In human medicine a significant loss of abdominal fat, the type associated with metabolic syndrome, has been reported in studies investigating capsinoid supplements. This effect may be particularly beneficial considering that central adiposity is a risk factor for cardiovascular disease.

> These properties are thought to occur through stimulation of the TRPV1 (transient receptor potential vanilloid type-1) receptor, a cation channel involved in the regulation of body temperature. Activation of the receptor causes a release of catecholamines that then stimulates the central nervous system via beta receptors. The ability of capsinoids to produce the thermogenic properties of capsaicinoids but lacking the pungency, relates to

> their inability to stimulate receptors in the mouth, due to structural differences. However, the appetite reducing effects of chilli peppers may be sensory in origin and relate to the ability of capsaicinoids to stimulate oral receptors. Therefore capsinoids may not be as effective as the capsinoids at reducing hunger and increasing satiety. Most weight loss studies to date have involved the use of supplements. However, liberal use of chillies and powder in food would likely have similar effects, although the capsinoid content of regular chillies is low (Barrington, 2014).

In addition to these thermogenic properties, the capsaicinoids may also increase satiety. Following ingestion of food containing capsaicin, *ad libitum* energy intakes decrease and ratings of hunger are reduced (Jansens et al., 2014).

> The exact mechanism of action is not fully understood; however, capsaicin has been proven to increase concentrations of the anorexigenic hormone glucagon-like peptide 1 and to decrease concentrations of the orexigenic hormone ghrelin (Smeets and Westerterp-Plantenga, 2009). Furthermore, a study in rats using a diet containing capsaicin has shown that this extract causes an increase in catecholamine secretion from the adrenal medulla caused by activation of the central nervous system (Watanabe et al., 1987). There is an association between sympathetic nervous system (SNS) activity and food intake behavior: food intake decreases when SNS activity increases (Russek et al., 1987). This increase in SNS activity after ingestion of capsaicin suggests that the reduction in energy intake could be due to the anorexigenic effect of catecholamines (Yoshioka et al., 1999).

Therefore there is plenty of evidence that the capsaicinoids and capsinoids have potential beneficial effects at weight loss in pets or even as a nutriaceutical measure preventing it. While capsaicinoids are not a magic bullet for weight loss, the evidence is that they could play a beneficial role, as part of a weight management program (Whiting et al., 2012). Further research is expected in this topic in the next years.

Capsaicin and Immunity

Capsaicin, as a plant extract is included in the group of terpenoids (Cowan, 1999). The antimicrobial mechanism of action of terpenes is not fully understood but is speculated to involve membrane disruption by the lipophilic compounds.

Several reports document the antimicrobial properties of capsaicin (Cichewicz and Thorpe, 1996), the ability of dietary capsaicin to enhance disease resistance in broiler chickens fed 20 ppm capsaicin, decreasing the incidence of *Salmonella* colonization of the ceca (Orndorff et al., 2005) and the inhibitory effects of capsaicin on nitric oxide and prostaglandin E2 (PGE2) synthesis (Chen et al., 2003). Other authors reported that dietary concentrations of 20 ppm capsaicin enhanced T-lymphocyte mitogen-induced proliferation, B-lymphocyte numbers, and the serum concentration of immunoglobulin G (IgG) and M (IgM) in mice (Yu et al., 1998). Collectively, these data show the potential of capsaicin to modulate various arms of the immune system as well as the response to an antigenic challenge.

Capsaicin could act as an antimicrobial agent against bacterial pathogens, for example, *Bacillus spp.* (Cichewicz and Thorpe, 1996). One recent study showed that the sub-bactericidal concentration of capsaicin drastically inhibits cholera toxin production in various serogroups of *Vibrio cholera* (Chatterjee et al., 2010). In rats, Qiu et al. (2012) demonstrated that capsaicin drastically suppressed the production of α-toxin by *Staphilococcus aureus* without affecting their viability, protected α-toxin-mediated alveolar epithelial cell injury *in vitro* and is also able to prevent the mortality associated with *S. aureus* pneumonia. The study observed the effects of three experimental subcutaneous doses of capsaicin (100, 50, or 25 mg of capsaicin per kg) two hours port-infection. Results showed that the mice that received 100 mg/kg of capsaicin were significantly protected against *S. aureus* pneumonia at 24 h, 48 h and 72 h. When the mice were administered 50 mg/kg of capsaicin, they were significantly protected from mortality 24 h post-infection. Finally, at a 25 mg/kg dose, no significant protected effects were observed. They concluded that capsaicin may potentially be useful for the treatment of *S. aureus* pneumonia when used in combination with commonly used antibiotics.

> Anti-virulence drugs such as capsaicin could be used in combination with established or novel antimicrobials in a synergistic manner to increase the clinical performance and extend the lifespan of these drugs (Cegelski et al., 2008; Paul and Leibovici, 2009).

On the contrary, Kamel (2000) observed that capsicum extract showed specific efficacy against *Clostridium spp.* alone and not against *E. coli H7:O157, S. typhimurium* or *E. coli*. For the author, these differences in efficacy at the *in vitro* level can be clearly linked to the level of their active substances, with purity of the plant extract playing a critical role.

Hardin et al. (2007) analyzed the effect of dietary capsaicin supplementation (6.75 ppm or 12.5 ppm/animal/day) on articular immune function in horses. The study concluded that dietary capsaicin, at the levels fed in the experiment did not appear to provide beneficial effects in terms of joint health or immune response (serum immunoglobulin G, haptoglobin and α-1 acid glycoprotein) after an antigenic challenge with tetanus toxoid. Authors attributed the lack of effects to the unknown mechanisms of absorption and metabolism of capsaicin before reaching the horses' systemic circulation.

> Previous reports have shown that capsaicin is not readily absorbed from the gut lumen in avians, leaving the intact capsaicin molecule to exert its antimicrobial effects on the cecal environment or microbial population (Tellez et al., 1993). However, in rats, capsaicin is readily absorbed from the gut lumen, but the molecule is metabolized in the liver before reaching the general circulation (Donnerer et al., 1990).
>
> Independently of the dosage, authors hypothesized that the dietary capsaicin was either not absorbed as an intact molecule or that the capsaicin was degraded before entering the systemic circulation, thereby preventing binding of capsaicin to its receptor on other cells throughout the body.

Demirbilek et al. (2004) used low and high doses of capsaicin on the modulation of systematic inflammatory response in different states of septicemia in rats. They demonstrated that a low dose of capsaicin increased the anti-inflammatory effect of interleukin 10 (IL-10) and decreased the proinflammatory cytokine, NOX (nitrite/nitrate), and tissue MAD (malondialdehyde) in the mice with septicemia. According to Spiller et al. (2008) capsaicin could block the inflammatory cytokine in the inflammation site.

It is well established that oxidative stress is involved in the etiology of different diseases and occurs also as a consequence of increased immune system activity (Castillo et al., 2001). Thus, an antioxidant support to sustain the DNA integrity of the immune cells may be beneficial for preserving healthy animals and the quality of animal products. In this line, Frankic et al. (2009) showed that supplementation with a mixture composed of carvacrol, capsicum oleoresin and cinnamaldehyde was effective against high n-3 polyunsaturated fatty acid (PUFA)-induced oxidative stress in young pigs.

> The n-3 PUFA-induced oxidative stress accelerates the formation of free radicals, which can also damage the DNA, resulting in single/double strand breaks or oxidized bases (Salobir et al., 2005).

In cats, the role of capsaicin has been studied in acute antigen-induced knee arthritis (Inman et al., 1989), concluding that animals treated with intra-articular capsaicin showed a diminution in the severity of inflammation compared with contralateral arthritic joint, moderating the severity of inflammation.

Chili is used worldwide for the treatment of internal parasites such as ovine fasciolosis. In Mexico, 13 chili peppers blended with water were used to treat ovine fasciolosis (McCorkle et al., 1995). Ahmed et al. (2007) compared the effect of chili and the traditional treatment such as tricalbendazole in the treatment of fasciolosis, at the end of the experiment tricalbendazole and chili treated groups of animals had an efficacy of 100% and 84%, respectively. They concluded that under subclinical conditions or as a preventive measure, chili supplements might be an alternative.

Capsaicin and Digestive System

Ruminal Homeostasis

Herbs and spices have been introduced also to ruminant nutrition. Microbial ecosystem in the rumen is composed from complex anaerobic microbial population of bacteria, fungi, protozoa, methanogeneous arhea and bacterifagi. Numerous metabolites produced in rumen during microbial fermentation affect the basic digestive and metabolic functions and productivity of the host. Researchers have been searching for new possibilities to modulate the microbial fermentation processes in the rumen. The main goal of manipulating the rumen fermentation is to increase the effectiveness of digestion and metabolism of nutrients, in order to increase the productivity of the animals and to suppress undesirable processes, such as methanogenesis. In intensive farming systems, feed additives, including antibiotics; were used to increase the milk, meat and wool production. The ban on antibiotic use in Europe increases the production costs what triggered the need to search for antibiotic alternatives in ruminant nutrition (Frankic et al., 2009).

Capsaicin, as component of capsicum oil (10 to 15%; Cichewicz and Thorpe, 1996) is currently used as a modifier of rumen microbial fermentation processes (Calsamiglia et al., 2007).

When supplied to an *in vitro* culture of rumen fluid obtained from dairy cattle, the effects have been negligible in the short and long term, when a 60% alfalfa hay and 40% concentrate diet was used (Cardozo et al., 2004; Busquet et al., 2005). However, Cardozo et al. (2005) demonstrated that the effects were different in an *in vitro* system with rumen fluid from beef cattle fed a 10:90 straw:concentrate ratio diet, and reported that at pH 7.0 total Volatile Fatty Acids (VFA) and ammonia-N concentrations were reduced and the acetate-to-propionate ratio increased. In contrast, at pH 5.5, capsicum oil reduced ammonia-N concentration, increased total VFA production and propionate proportion and reduced acetate proportion and the acetate-to-propionate ratio.

> Therefore, its effects on high concentrate diets at low pH suggest that nutrient utilization in the rumen may be improved. The low pH appears to shift the molecule to a more hydrophobic status, being more active as an antimicrobial.

Cardozo et al. (2006) tested the effect of feeding capsicum oil in rumen-cannulated beef cattle fed a high concentrate diet. Total VFA concentration was not affected, but the molar proportion of acetate decreased. Capsicum oil reduced the ruminal concentration of large peptides and increased that of small peptides and amino acids, but had no effect on ammonia-N concentrations.

> These effects suggest that it stimulates peptidolysis, which may provide more peptides and amino acids, and enhance microbial protein synthesis and flow to the small intestine. Although these changes were in the same direction as in the *in vitro* study of Cardozo et al. (2005), effects were smaller. However, this effect was not always consistent (Cardozo et al. 2004; Busquet et al., 2005).

Nevertheless, Calsamiglia et al. (2007) pointed out the potential effects of capsaicin on rumen microbial fermentation altering microbial N-metabolism. The decrease on ruminal NH_3 concentration suggested that the main mechanism of action is through the inhibition of deamination.

> Foskolos (2012) investigated the effect of capsaicin in the optimization of N-use in the rumen and the role played by this result in the reduction of environmental impact of dairy production. To study this, the authors prepared and treated silage with capsaicin –among other extracts- at three different concentrations: 50, 500 and 2000 mg/kg. Curiously, authors observed no

significant effects of capsaicin on silage chemical characteristics and protein degradation, with other extracts, such as eugenol, thymol, carvacrol or cinnamaldehyde, providing better results.

Usually, capsaicin is given in combination with other plant extracts favoring the synergy among them. One example is the mixture of capsaicin, cinnamaldehyde and eugenol; in this mix the capsaicin also shows the effects of the other two components, but in addition also modifies the distribution of feed consumption over the course of a day. It reduces the size of the first meal after feed is offered and increases consumption at further visits to the feedbunk, reducing acidosis risk resulting from the first big feed intake of the day (Vikari, 2008).

When capsicum was tested *in vivo* with feedlot heifers, DMI and water intake consistently increased (Cardozo et al., 2006; Fandiño et al., 2008). However, the increase in DMI occurred without affecting rumen pH. In this sense, for Rodriguez-Prado et al. (2012) capsicum, because of its pungent effect, stimulated water consumption, DMI, and the number of food intakes throughout a day (see Table).

With this aim, four beef heifers were fed four different treatments: control (no additives), 125 (CAP125), 250 (CAP250), and 500 (CAP500) mg/day of capsicum oleoresin standardized with 6% of capsaicin and dihydrocapsaicin. Dry matter intake in the first 2 h after feeding was significantly reduced in all CAP treatments compared with the control group. After this adaptation period, a linear response was observed (CTR, CAP125, CAP250, and CAP500, respectively) for DMI (8.56, 9.84, 8.68, and 9.40 kg/d, respectively).

The molar proportion of acetate tended to decrease linearly (from 59.6 to 55.5 mol/100 mol), and ammonia N concentration tended to increase linearly (from 14.4 to 16.0 mg N/dL). In contrast, the molar proportion of propionate (23.8 mol/100 mol), butyrate (14.2 mol/100 mol), and lactate (0.28 mol/100 mol) were not affected by supplementation with capsaicin (see Table in this position). These results indicated that capsicum extract stimulated DMI and modified the pattern of DMI in beef cattle fed high concentrate diets.

Table 2. Effects of increasing dose of capsaicin extract on mean values of ruminal pH and fermentation characteristics of beef heifers fed a high-concentrate diet (from Rodriguez-Prado et al., 2012)

	Treatments[1]					P value		
Item	CTR	CAP125	CAP250	CAP500	SEM	Linear	Quadratic	T(H)[2]
Rumen pH	6.03	5.84	5.96	5.86	0.089	0.08	0.43	0.63
N-NH3 (mg/100mL)	14.4	14.4	15.2	16.0	3.46	0.08	0.80	0.89
Total VFA mM	134.3	144.8	140.1	142.8	2.71	0.08	0.14	0.03
Ac:Pr[3] Mol/100mL	3.18	2.76	3.25	2.39	0.439	0.12	0.46	0.96
Acetate	59.6	59.5	59.9	55.5	2.47	0.06	0.26	0.87
Propionate	22.3	24.2	21.8	27.2	4.14	0.25	0.53	0.98
Butyrate	14.8	13.6	14.6	13.5	1.85	0.60	0.98	0.92
Valerate	1.82	1.74	1.79	1.92	0.073	0.31	0.16	0.46
Lactate	0.27	0.19	0.18	0.18	0.044	0.26	0.29	0.56
BCFA[4]	1.73	1.68	1.83	1.70	0.080	0.98	0.49	0.42

[1]Treatment were CTR = no additive; CAP125 = capsicum extract 125 mg/animal/day; CAP250 = capsicum extract 250 mg/animal/day; CAP500 = capsicum extract 250 mg/animal/day.

[2]Treatment (hour) effect.

[3]Acetate to propionate ratio.

[4]BCFA = Branched-chain fatty acids (isobutyrate + isovalerate).

Digestion in Monogastrics

The first step in traceability is in the determination of what level of the compound crosses the intestinal epithelium. If the substance in question and its respective metabolites remain intraluminal and are finally excreted in the urine or feces, there is little need to check for its deposition in internal depots (Kamel, 2000).

> In this regard, immortalized epithelium cell systems, such as the Caco-2 cell line -derived from a human colon adenocarcinoma- are considered the most similar morphologically and functionally to intestinal epithelial cells. Their use has been quite extensive in the area of pharmacology, especially since there is a high degree of correlation between the permeability to pharmacological agents and the absorption of these same compounds in live subjects. This demonstrates only one value of the system in predicting potential bioavailability and absorbability of active compounds.

> Another tool, representative of the *in vitro* systems being developed in the public sector, is the monogastric intestinal model, developed by Drs. Robert Haavenar and Mans Minekus at the TNO Institute in Zeist, Netherlands. This "mechanical" model provides continuous monitoring of the natural parameters of digestion and its by-products. This can in turn determine the traceability of ingested feed components as well as their possible effects on the animal and their impact on the environment. Briefly, the TNO *in vitro* model is composed by 4 successive compartments simulating the stomach, duodenum, jejunum, and ileum (Minekus et al., 1995).

Capsicum oleoresin (*Capsicum annum*) has been considered as a digestion stimulant (Frankic et al., 2009). In rats, dietary capsaicin (15 mg % MS), enhanced intestinal lipase activity and also the disaccharidases sucrase and maltase contributing to stimulate digestion (Platel and Srinivasan, 1996). Later Platel and Srinivasan (2000) demonstrated that the dietary intake of capsaicin prominently enhanced the activity of lipase (around 30–37%), the activity of amylase (67%) the activity of trypsin (more than 120%) and the activity of chymotrypsin; all of these enzymes in comparison with control rats (Table 2).

> Pancreatic lipase plays a vital role in the digestion of fat. The significant enhancing influence of the spice active ingredient capsaicin on pancreatic lipase suggests that this spice ingredient has a beneficial stimulatory effect on digestion.

But capsaicin not only stimulates pancreatic lipase secretion but also intestinal lipase. This enzyme, which is distinct from pancreatic lipase, has been claimed to significantly supplement the pancreatic enzyme in the hydrolysis of dietary fat, specifically monoglycerides. Intestinal lipase probably assumes a greater role in the digestion of fat when the titers of pancreatic lipase are limiting.

On the other hand, a higher rate of biliary secretion of bile acids has been documented in rats fed capsaicin (Ganesh Bhat et al. (1984 and 1985), quoted by Platel and Srinivasan, 2000).

Table 3. Effect of dietary spices on the activities of lipase, amylase, trypsin and chymotrypsin of rat pancreas (from Platel and Srinivasan, 2000). Values are mean ± SEM of 12 animals/group. The test spices were administered intragastrically to overnight fasted animals as 1 mL suspension in 10% starch solution at the following doses: curcumin, 100 mg; capsaicin, 3 mg; piperine, 10 mg; ginger (fresh). A parallel group of control rats received 1 ml 10% starch solution

	Enzymes			
Diet group	Lipase[1]	Amylase[2]	Trypsin[3]	Chymotripsin[3]
Control	30.5±1.85	47.0±2.5	2.59±0.13	2.89±0.05
Curcumin	54.8±3.91*	92.1±9.9*	6.59±0.67*	5.0±0.28*
Capsaicin	41.4±1.81*	80.9±8.0*	5.70±0.71*	3.62±0.11*
Piperine	41.8±2.76*	88.0±11.8*	6.87±0.52*	3.75±0.27*
Ginger	39.4±3.34*	138.0±10.0*	6.03±0.71*	3.75±0.14*

[1]μmol free fatty acids liberated/(min x mg protein).
[2]nmol maltose liberated/(min x mg protein).
[3]nmol p-nitroaniline released/(min x mg protein).
[*]Significantly higher than corresponding control value ($P < 0.05$).

> The role of bile and biliary bile acids in fat digestion and absorption is well known. Hence, stimulation of bile secretion, especially that of bile with higher levels of bile salts is probably a major mechanism by which this spice aids in digestion.

Other properties of capsaicine have been described such as to induce a greater retention time of the diet (Platel et al., 2002), and to increase gastrointestinal blood flow (Leung, 1993). This increase in the retention time and irrigation has been proposed to improve HCl secretion and nutrient digestion and absorption (Dunshea, 2003).

> Improved digestive capacity in the small intestine may be considered as an indirect side effect of phytogenic stabilization of the microbial eubiosis in the gastrointestinal tract. Consequently, phytogenic relieve the host animals from immune defense stress during critical situations to increase the intestinal availability of essential nutrients for absorption; thereby helping animals to grow better within the framework of their genetic potential (Windisch et al., 2008 quoted by Hashemi and Davoodi, 2011).

As we have seen throughout this chapter, capsaicin has complex physiological effects, eliciting both pain and producing analgesia, both mucosal protection and injury. In light of the complex and seemingly conflicting actions, capsaicin has been referred to as a `double-edged sword' (Rodriguez-Stanley et al., 2000). Studies performed in human beings have demonstrated that capsaicin seems to protect oesophageal and gastric mucosa trough increasing regional blood flow (Bass et al., 1991; Kang et al., 1996; Teng et al., 1998).

> The study performed by Holzer and Sametz (1986) demonstrated that capsaicin-sensitive afferent neurons are involved in gastric mucosal protection against ulcerogenic factors. Their data further suggested that this type of gastric defense was primarily due to a local mechanism initiated by sensory nerve endings in the gastric mucosa.

Nevertheless, the therapeutic application of this active ingredient in the treatment of gastric disorders in veterinary medicine needs further studies.

Natural capsaicin is available in liquid form as capsicum oleoresin, a highly irritating product that has limited direct usage as an in-feed additive for livestock diets. Controlling the biopharmaceutical behavior of capsaicin in the digestive environment is necessary to ensure its optimum use. Although spray-drying techniques afford some protection for capsaicin, the high water solubility values of the carriers used cannot control its release into the gastrointestinal tract. For this reason, the search of new techniques is continuously required. Meunier et al. (2007) proposed the microencapsulation produced by spray-cooling technology as a technique able to decrease the irritant effect of this oleoresin and to control its release throughout the gastrointestinal tract in a cost-effective way (Figure 6).

> Capsaicin can be obtained directly from ground spices, but the use of the oleoresin form is more hygienic, easily standardizable, more concentrated in capsaicin, and thus requires less storage space. However, oleoresin is a highly irritant product, which limits its direct use as a micro-additive for livestock diets and, in this liquid form, is sensitive to light, heat, and oxygen, thereby having a short storage life if not stored properly. Microencapsulation protects oleoresin against such destructive changes and also converts it into a free-flowing powder that can be easily and homogeneously mixed with feed, which makes administration easier. Moreover, microencapsulation can control the biopharmaceutical behavior of the additive in the gastrointestinal tract (Meunier et al., 2007).

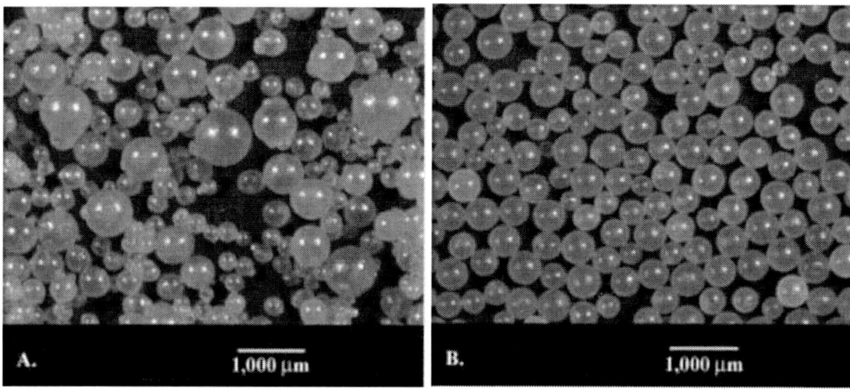

Figure 6. Stereomicroscopy of microsphere (20% of capsicum oleoresin into a rapeseed, hydrogenated oil matrix (from Meunier et al., 2007). Microencapsulation: Two technologies of spray cooling were used to produce the microspheres, a fluidized air bed using a spray nozzle (A) and a system with a vibrating nozzle placed on top of a 5-m cooling tower (B). Although the vibrating nozzle technology allowed better control of the particle size than the spray nozzle, in opinion of the authors both spray-cooling techniques gave satisfactory results for the general aspect of the microspheres.

Hardin et al., (2007) observed that horses required approximately 2 days to become used to consuming the daily ration containing capsaicin, although this response depended strongly on the animal: some horses had no aversion to the beet pulp-capsaicin mixture and others refused to eat the entire daily ration.

> This was overcome by mixing some of the daily alfalfa ration into the beet pulp capsaicin mixture to ensure that each horse consumed its daily dose of capsaicin. The predominant behavioral change noted was excess salivation in some horses and increased wood chewing of the stalls.

CAPSAICIN AS A GROWTH PROMOTANT

Feed supplements with growth promoting activity increase stability of feed and beneficially influence the gastrointestinal ecosystem, mostly through growth inhibition of pathogenic microorganisms' growth. Due to improved health status of the digestive system, animals are less exposed to toxins of microbiological origin. Consequently, herbs and spices help to increase the resistance of the animals exposed to different stress situations and increase the absorption of essential nutrients, thereby improving the growth of the animals (Windisch et al., 2008; Frankic et al., 2009).

> Castillo et al. (2006) reported that the mixture of cinnamaldehyde, capsicum oleoresin and carvacrol enhanced the growth of *Lactobacilli*, increasing the ratio of *Lactobacilli* to *Enterobacteria,* contributing to modulate the composition of microbial population by prebiotic activity.

The role of capsaicin in this topic has been considered mainly in swine. Thus, Dunshea et al. (2003) demonstrated an improvement in growth performance with the inclusion of vanillylnonamide, a capsaicin analogue, in the diets of finisher pigs. On the contrary, Frankic et al. (2009) did not observe significant improvements in body weight of pigs supplemented with a mix of plant extracts were capsaicin was present and independently of its positive effect on the redox balance. However, the combination of carvacrol, cinnamaldehyde and capsicum oleoresin beneficially affected gastrointestinal ecosystem and gastric emptying of weaned pigs (Manzanilla et al., 2004).

Studies like the one by Meunier et al. (2007), considered capsaicin as a valid growth promotant in pigs as alternative to in-feed antibiotics. Nevertheless, for these authors the main problem was the way of administration, due to the pungent properties of the plant extract. Microencapsulation of capsicum oleoresin could reduce the irritating effect of capsicum oleoresin and help controlling its release kinetics during consumption by the animal

In the study of Hardin et al. (2007) horses were divided in three groups depending on the dosage of capsaicin in diet: control; control diet plus 50 mg capsaicin/head/d (CAP50); and control diet plus 100 mg capsaicin/head/day (CAP100). The CAP50 group received 6.75 ppm and the CAP100 one, 12.5 ppm capsaicin in the total diet, respectively. Their results showed that body weight of horses increased significantly over the course of the trial in comparison with controls, but did not differ between capsaicin treatments.

In ruminants, capsaicin has been given in combination with other plant extracts. A field study performed by Hernández et al. (2009) investigated the effect of using of a blend of capsaicin, cinnamaldehyde and carvacrol supplementation on beef cattle production, finding a positive effect of supplementation on production parameters, especially during the finishing period. On the other hand, serum L-lactate (an indirect biomarker of ruminal acidosis), was decreased in the supplemented animals. In relation to other blood metabolites, this supplement did not produce any detrimental effect, but the contrary was found: the decreased NEFA and creatinine levels, along with the increase in serum urea nitrogen concentrations; suggested an improvement in the energy status and protein turnover of the supplemented animals (Castillo et al., 2012).

Petkova et al. (2008) supplemented dairy cows with a nutritional additive containing methyl-vanillyl-nonenamide-capsaicin trying to study the possible effects of this additive on the blood lipids metabolism in lactating cows. Their results showed that there were not changes in feed intake between supplemented and no supplemented cows. However, the positive effects observed in lipid profile (decrease in LDL and VLDL cholesterol as well as triglycerides and total lipids) make this additive suitable to use as a feed additive in lactating cows.

> Other positive effects observed by the above-mentioned additive were on the laying rate in pheasants (Kitanov et al., 2000b), on reproductivity of hens (Kitanov et al., 2000a), turkeys (Kitanov et al., 2003), ewes (Djorbineva et al., 2006) and cows (Petkova et al., 2008) and finally on the percentage of milk fat in sheep (Djorbineva et al., 2006).

Hernández et al. (2004) showed that a blend of carvacrol, cinnamaldehyde and capsaicin improved feed digestibility in broilers. The authors attributed the positive effects of plant extracts on nutrient digestibility to the appetite and digestion-stimulating properties and antimicrobial effects.

Since a few years, literature is abundant in relation to the positive effects of capsaicin or its derivate substances, in combination with other plant extracts on promoting growth in beef cattle as a substitute of monensin in a high concentrate diet (Fandiño et al., 2008; Geraci et al., 2012; Rodriguez-Prado et al., 2008). Thus, further research in this item is expected in the upcoming years.

In conclusion, there are many advantages of using capsaicin as a natriucetical or therapeutic tool in Veterinary Medicine. The future of using capsaicin will depend, to a great extent, on the knowledge that we have about its metabolic pathways in the different species, and the technological measures that could facilitate its use as both a nutrient and a therapeutic agent.

REFERENCES

American College of Toxicology (2007). Final report on the safety assessment of *Capsicum annuum extract*, *Capsicum annuum* fruit extract, *Capsicum annuum* resin, *Capsicum annuum* fruit powder, *Capsicum frutescens* fruit, *Capsicum frutescens* fruit extract, *Capsicum frutescens* resin, and capsaicin. *International Journal of Toxicology* 26: 3-105.

Anand P, Bley K (2011). Topical Capsaicin for Pain Management. Therapeutic Potential and Mechanisms of Action of the New High-concentration Capsaicin 8% Patch. British Journal of Anaesthesia 107(4): 490-502.

Anderson WS, Sheth RN, Bencherif B, Frost JJ, Campbell JN (2002). Naloxone increases pain induced by topical capsaicin in healthy human volunteers. *Pain* 99: 207–216.

Barrington R (2014). Capsaicinoids and Capsinoids. Available at: http://www.robertbarrington.net/capsaicinoids-and-capsinoids

Bass BL, Trad KS, Harmon JW, Hakki FZ (1991). Capsaicin-sensitive nerves mediate esophageal mucosal protection. *Surgery* 110: 419-425.

Bennett DJ, Kirby GW (1968). Constitution and biosynthesis of capsaicin. *Journal of Chemistry. Society* 4:442–446

Bley KR (2004). Recent developments in transient receptor potential vanilloid receptor 1 agonist-based therapies. *Expert Opinion on Investigational Drugs* 13:1445–1456.

Börszei R. (2012). Investigation of the activation mechanisms of capsaicin-sensitive sensory nerve terminals and measurement of the released neuropeptides. *PhD Thesis,* University of Pécs, Faculty of Medicine, Hungary.

Busquet M, Calsamiglia S, Ferret A, Kamel C (2005a). Screening for the effects of natural plant extracts and secondary plant metabolites on rumen microbial fermentation in continuous culture. *Animal Feed Science and Technology* 123/124: 597-613.

Calsamiglia S, Busquet M, Cardozo PW, Castillejos L, Ferret A, Fandiño I (2007). The use of essential oils in ruminants as modifiers of rumen microbial fermentation. 2007-Penn State Dairy Cattle Nutrition Workshop. Available at: http://extension.psu.edu/animals/dairy/courses/dairy-cattle-nutrition-workshop

Cardozo PW, Calsamiglia S, Ferret A, Kamel C (2004). Effects of natural plant extracts on protein degradation and fermentation profiles in continuous culture. *Journal of Animal Science* 82: 3230-3236

Cardozo PW, Calsamiglia S, Ferret A, Kamel C (2005). Screening for the effects of natural plant extracts at different pH on in vitro rumen microbial fermentation of a high-concentrate diet for beef cattle. *Journal of Animal Science* 83: 2572-2579

Cardozo PW, Calsamiglia S, Ferret A, Kamel C (2006). Effects of alfalfa extract, anise, capsicum and a mixture of cinnamaldehyde and eugenol on ruminal fermentation and protein degradation in beef heifers fed a high concentrate diet. *Journal Animal Science* 84: 2801-2808.

Castillo C, Benedito JL, Lopez-Alonso M, Miranda M, Hernandez J (2001). Importance of oxidative stress in cattle: its relationship with the physiological condition (pregnancy and parturition) and nutrition. *Archivos de Medicina Veterinaria* 33 (1):5-20.

Castillo C, Benedito JL, Vázquez P, Pereira V, Méndez J, Sotillo J, Hernández J (2012). Effects of supplementation with plant extract product containing carvacrol, cinnamaldehyde and capsaicin on serum metabolites and enzymes during the finishing phase of feedlot-fed bull calves. *Animal Feed Science and Technology* 171: 246–250.

Castillo M, Martín-Orúe SM, Roca M, Manzanilla EG, Badiola I, Perez JF, Gasa J (2006). The response of gastrointestinal microbiota to avilamycin, butyrate, and plant extracts in early-weaned pigs. *Journal of Animal Science* 84: 2725–2734

Cegelski L, Marshall GR, Eldridge GR, Hultgren SJ (2008) The biology and future prospects of antivirulence therapies. *Nature Reviews Microbiology* 6: 17–27.

Chanda S, Bashir M, Babbar S, Koganti A, and Bley K (2008). In Vitro Hepatic and Skin Metabolism of Capsaicin. *Drug Metabolism and Disposition* 36(4): 670-675

Chatterjee S, Asakura M, Chowdhury N, Neogi SB, Sugimoto N, Haldar S, Awasthi SP, Hinenoya A, Aoki S, Yamasaki S (2010). Capsaicin, a potential inhibitor of cholera toxin production in *Vibrio cholerae*. *FEMS Microbiology Letters* 306: 54–60.

Chen C, Lee S, Wu W, Fu W, Ho F, Lin W. (2003). Signal transduction for inhibition of inducible nitric oxide synthase and cyclooxygenase-2 induction by capsaicin and related analogs in macrophages. *British Journal Pharmacology* 140: 1077–1087.

Chhabra N, Aseri ML, Goyal V, Sankhla S (2012). Capsaicin: A promising therapy-A critical reappraisal. *International Journal of Nutrition, Pharmacology, Neurological Diseases*. 2(1): 8-15.

Cichewicz R, Thorpe P (1996). The antimicrobial properties of chile peppers (Capsicum species) and their uses in Mayan medicine. *Journal Ethnopharmacology* 52: 61–70.

Cowan MM (1999). Plant products as antimicrobial agents. *Clinical Microbiology Reviews* 12(4): 564–582.

Degim IT, Uslu A, Hadgraft J, Atay T, Akay C, Cevheroglu S (1999). The effects of Azone and capsaicin on the permeation of naproxen through human skin. *International Journal Pharmacology* 179: 21–25.

Demirbilek S, Ersoy MO, Demirbilek S, Karaman A, Gürbüz N, Bayraktar N, Bayraktar M (2004). Small-dose capsaicin reduces systemic inflammatory responses in septic rats. Anesthesia and Analgesia 99: 1501-1507.

Di Giminiani P, Petersen LJ, Herskin MS (2014). Capsaicin-induced neurogenic inflammation in pig skin: A behavioural study. *Research in Veterinary Science* 96: 447-453.

Djorbineva M, Kitanov I, Mihailova G, Dimitrov I (2006). The effect of supplemental feeding by "*Ovocap*©" over some milk and reproductive characteristics in dairy sheep. *Proceedings of scientific conference The science under the globalization conditions*, Stara Zagora, 2, 327-332.

Donnerer J, Amann R, Schuligoi R, Lembeck F (1990). Absorption and metabolism of capsaicinoids following intragastric administration in rats. *Naunyn-Schmiedebergs Archives of Pharmacology* 342: 357-361.

Dray A (1992) Neuropharmacological mechanism of capsaicin and related substances. *Biochemical Pharmacology* 44: 611-615.

Dunshea FR (2003). Metabolic and endocrine changes around weaning. In: Pluske JR, Le Dividich J, Verstegen MW (Eds). *Weaning the Pig. Concepts and Consequence.* Wageningen Academic Publishers, Wageningen, the Netherlands.

Dunshea FR, Suster D, Kerton DJ, Leury BJ (2003). A capsacin analogue improves growth and dressing rate in pigs, particularly gilts. Paterson JE (Ed.) *Manipulating pig production*. Australasian Pig Science Association Inc., Werribee, Australia.

Endris Feki Ahmed EF, Markvichitr K, Tumwasorn S, Koonawootrittriron S, Choothesa A, Jittapalapong S (2007). Anthelminthic efficacy of chili (*Capsicum annum longum*) and nutritional supplementation on ovine fascioliasis in Ethiopia. *Kasetsart Journal (Natural Science)* 41:724-732.

Fandiño I, Calsamiglia S, Ferret A, Blanch M (2008). Anise and capsicum as alternatives to monensin to modify rumen fermentation in beef heifers fed a high concentrate diet. *Animal Feed Science and Technology* 145: 409–417.

Fang JY, Wu PC, Huang YB, Tsai YH (1995). *In vitro* permeation study of capsaicin and its synthetic derivatives from ointment bases using various skin types. *International Journal of Pharmacology* 126: 119–128.

Foskolos A (2012). Strategies to reduce nitrogen excretion from ruminants: targeting the rumen. *PhD Thesis*, Universidad Autónoma de Barcelona, Facultad de Veterinaria, España.

Frankic T, Levart A, Salobir J (2009). The effect of vitamin E and plant extract mixture composed of carvacrol, cinnamaldehyde and capsaicin on oxidative stress induced by high PUFA load in young pigs. *Animal* 4 (4): 572–578.

Fusco BM, Giacovasso M (1997). Peppers and pain: The promise of capsaicin. *Drugs* 53: 909–914.

Geraci JI, Garciarena AD, Gagliostro GA, Beauchemin KA, Colombatto D (2012). Plant extracts containing cinnamaldehyde, eugenol and capsicum oleoresin added to feedlot cattle diets: Ruminal environment, short term intake pattern and animal performance. *Animal Feed Science and Technology* 176 (1-4): 123–130.

German AJ (2006). The growing problem of obesity in dogs and cats. Journal of Nutrition 136: 1941S-1946S.

Greathead H (2003). Plant extracts for improving animal productivity. *Proceedings of the Nutrition Society* 62: 279-290.

Greaves M, Wall PD (1996). Pathophysiology of itching. *Lancet* 348: 938–940.

Hardin S, Potts L, Waggoner J, Loëst C, Turner J (2007). Dietary capsaicin does not alter synovial concentrations of prostaglandin E2 or the acute phase response in aged horses after antigenic challenge. *Journal of Equine Veterinary Science* 27 (11): 494-498.

Hargreaves KM, Jackson DL, Bowles WR (2003). Adrenergic regulation of capsaicin-sensitive neurons in dental pulp. J Endodontics 29: 397-399.

Hashemi SR, Davoodi H (2011). Herbal plants and their derivatives as growth and health promoters in animal nutrition. *Veterinary Research Communications* 35:169–180.

Hernandez F, Madrid J, Garcia V, Orengo J, Megias MD (2004). Influence of two plant extracts on broiler performance, digestibility, and digestive organ size. *Poultry Science* 83:169–174.

Hernandez J, Benedito JL, Vazquez P, Pereira V, Méndez J, Sotillo J, Castillo C (2009). Supplementation with plant extracts (carvacrol, cinnamaldehyde and capsaicin): its effects on acid-base status and productive performance in growing/finishing bull calves. *Berliner und Münchener Tierarztliche Wochenschrift* 122 (3/4): 93-99.

Holzer P, Sametz W (1986). Gastric mucosal protection against ulcerogenic factors in the rat mediated by capsaicin-sensitive afferent neurons. *Gastroenterology* 91(4): 975-981

Inman RD, Chiu B, Rabinovich S, Marshall W (1989). Neuromodulation of synovitis: capsaicin effect on severity of experimental arthritis. *Journal of Neuroimmunology* 24 (1–2): 17-22.

Janssens PLHR, Hursel R, Westerterp-Plantenga MS (2014). Capsaicin increases sensation of fullness in energy balance, and decreases desire to eat after dinner in negative energy balance. *Appetite* 77C: 44–49.

Kamel C (2000). A novel look at a classic approach of plant extracts. *Feed Mix*, Special-2000, pp: 19-21.

Kang JY, Teng CH, Chen FC (1996). Effect of capsaicin and cimetidine on the healing of acetic acid induced gastric ulceration in the rat. *Gut* 38: 832-836.

Kathy K. Seino KK, Foreman JH, Greene SA, Goetz TE, Benson JG (2003). Effects of Topical Perineural Capsaicin in a Reversible Model of Equine Foot Lameness. *Journal of Veterinary Internal Medicine* 17: 563–566.

Kawada T, Iwai K (1985). *In vivo* and *in vitro* metabolism of dihydrocapsaicin, a pungent principle of hot pepper, in rats. *Agricultural and Biological Chemistry* 49: 441–448.

Kennis P (1982). Responses of single nerve fibers to capsaicin applied to the skin. *Neuroscience Letters* 29: 83–88.

Kitanov I, Dimitrov D, Kitanova G (2000a). Biological and histological studies in hens treated by *Ovocap.*© *Veterinary Medicine* 4 (4): 82-86.

Kitanov I, Nikolov B, Tiufekchiev K, Kitanova G (2000b). The *Ovocap*©– ecological feed supplement to active the reproduction in pheasants. *LTU, Jubilee Book Proceedings,* Sofia, Bulgaria.

Kitanov I, Oblakova M, Lalev M (2003). Effect of the *Ovocap*© preparation on reproductive parameters of hens and turkeys. *Bulgarian Journal Agricultural Science* 9: 521-526.

Kolonin MG, Saha PK, Chan L, Pasquani R, Arap W (2004). Reversal of obesity by targeted ablation of adipose tissue. *Nature Medicine* 10:625–32.

Krajewska AM, Powers JJ (1987). Gas chromatographic determination of capsaicinoids in green *Capsicum* fruits. *Journal AOAC* 70: 926–928.

Leung FW (1993). Inhibition of spinal afferent nerve-mediated gastric hyperemia by nicotine: Role of ganglionic blockade. *American Journal of Physiology* 264: H1087–H1092.

Liu M, Pertl C, Markowitz K, Dörscher-Kim J, Kim S (1992).The effects of capsaicin on pulpal blood flow. *Proceedings of the Finnish Dental Society* 88 (1): 463-467.

Lundblad L, Lundberg JM, Anggard A, Zetterstrom O (1987). Capsaicin sensitive nerves and the cutaneous allergy in man. *Allergy* 42: 20–25.

Madani Z, Seyedmajidi M, Moghadamnia A, Bijani A, Zahedpasha A (2012). Histologic evaluation of pulpal response to MTA and capsaicin in cats. *Caspian Journal of Dental Research* 1(1): 8-13.

Manzanilla EG, Perez JF, Martin M, Kamel C, Baucells F, Gasa J (2004). Effect of plant extracts and formic acid on the intestinal equilibrium of early-weaned pigs. *Journal of Animal Science* 82: 3210–3218.

Marsella R, Nicklin CF, Melloy C (2002). The effects of capsaicin topical therapy in dogs with atopic dermatitis: a randomized, double-blinded, placebo-controlled, cross-over clinical trial. *Veterinary Dermatology* 13: 131–139.

McCorkle CM, Mathias-Mundy E, Schillhorn van Veen TW (1995). Sheep husbandry and health care among Tozotzil Maya sheepherders. In: Perezgrovas R (Ed.) *Ethno-Veterinary Research and Development.* Available at: http://fadr.msu.ru/rodale/agsieve/txt/Vol6/4/art3.html

Meunier JP, Cardot JM, Manzanilla EG, Wysshaar M, Alric M (2007). Use of spray-cooling technology for development of microencapsulated capsicum oleoresin for the growing pig as an alternative to in-feed antibiotics: A study of release using in vitro models. *Journal of Animal Science* 85:2699-2710.

Minekus M, Marteau P, Havenaar R, Huis in 't Veld JHJ (1995). A multicompartmental dynamic computer-controlled model simulating the stomach and small intestine. *ATLA* 23:197–209.

Molyneux GS, Haller CJ, Mogg K, Pollitt CC (1994). The structure, innervation, and location of arteriovenous anastomoses in the equine foot. *Equine Veterinary Journal* 26: 305–312.

Nixon AJ, Cummings JF (1994). Substance P immunohistochemical study of the sensory innervation of normal subchondral bone in the equine metacarpophalangeal joint. *American Journal of Veterinary Research* 55: 28–33

Orndorff B, Novak C, Pierson F, Caldwell D, McElroy A (2005). Comparison of prophylactic or therapeutic dietary administration of capsaicin for reduction of *Salmonella* in broiler chickens. *Avian Disease* 49: 527–533.

Paul M, Leibovici L (2009). Combination antimicrobial treatment versus monotherapy: the contribution of meta-analyses. *Infectious Disease Clinics of North America* 23: 277–293.

Petersen LJ, Lyngholm AM, Arendt-Nielsen L (2010). A novel model of inflammatory pain in human skin involving topical application of sodium lauryl sulfate. *Inflammation Research* 59: 775-781.

Petkova M, Kitanov I, Girginov D (2008). Blood lipids profile in lactating cows fed with supplement of *Ovocap*©. *Biotechnology in Animal Husbandry* 24 (3-4): 19-28

Petkova M, Kitanov I, Zhelev K (2008). Effect of *Ovocap*© on the cows reproduction. *Journal of Animal Science* XLV(3): 132-137

Platel K, Rao A, Saraswathi G, Srinivasan K (2002). Digestive stimulant action of three Indian spice mixes in experimental rats. *Nahrung* 46: 394–398.

Qiu J, Niu X, Wang J, Xing Y, Leng B, Dong J, Li H, Luo M (2012). Capsaicin protects mice from community-associated methicillin-resistant *Staphylococcus aureus* pneumonia. *PLoS ONE* 7(3): e33032. doi:10.1371/journal.pone.0033032.

Reilly CA, Yost GS (2006). Metabolism of capsaicinoids by P450 enzymes: a review of recent findings on reaction mechanisms, bio-activation, and detoxification processes. *Drug Metabolism Review* 38: 685–706.

Reyes-Escogido ML, Gonzalez-Mondragon EG, Vazquez-Tzompantzi E. (2011). Chemical and Pharmacological Aspects of Capsaicin. *Molecules* 16: 1253-1270.

Rodríguez-Prado M, Ferret A, Zwieten J, Gonzalez L, Bravo D, Calsamiglia S (2012). Effects of dietary addition of capsicum extract on intake, water consumption, and rumen fermentation of fattening heifers fed a high-concentrate diet. *Journal of Animal Science* 90:1879-1884.

Rodriguez-Stanley S, Collings KL, Robinson M, Owen W, Miner PB (2000). The effects of capsaicin on reflux, gastric emptying and dyspepsia. *Alimentary Pharmacology and Therapeutics* 14: 129-134.

Russek M, Vega C, Barrera J, Soto-Mora LM, Lanzagorta A, Racotta R (1987). Anorexia elicited by different catecholamines in rats. *Appetite* 9 (2): 119–126.

Salobir J, Rezar V, Pajk T, Levart A (2005). Nucleotide supplementation effectively eliminates lymphocyte DNA damage induced by dietary oxidative stress in pigs. *Animal Science* 81: 135–140.

Smeets AJ, Westerterp-Plantenga MS (2009). The acute effects of a lunch containing capsaicin on energy and substrate utilisation, hormones, and satiety. *European Journal of Nutrition* 48 (4): 229-234.

Spiller F, Alves MK, Vieira SM, Carvalho TA, Leite CE, Lunardelli A, Poloni JA, Cunha FQ, de Oliveira JR (2008). Anti-inflammatory effects of red pepper (*Capsicum baccatum*) on carrageenan-and antigen-induced inflammation. *Journal of Pharmacy and Pharmacology* 60: 473-478.

Surh YJ, Lee SS (1995). Capsaicin, a double edged sword: toxicity, metabolism and chemoprotective potential. *Life Sciences* 56 (22): 1845-1855

Szallasi A, Blumberg PM (1999). Vanilloid (capsaicin) receptors and mechanisms. *Pharmacological Reviews* 51: 159–212.

Talati A, Akbari HA (2003). comparison of the effect of local application of capsaicin, zostrix and morphine to complete anesthesia of teeth with inflamed pulp and poor response to traditional local anesthetics. *Journal of Mashhad Dental School* 27: 31-36. [In Persian]

Tellez GI, Jaeger L, Dean CE, Corrier DE, DeLoach JR, Williams JD, Hargis BM (1993). Effect of prolonged administration of dietary capsaicin on *Salmonella enteritidis* infection in Leghorn chicks. *Avian Diseases* 37:143–148.

Teng CH, Kang JY, Wee A, Lee KO (1998). Protective action of capsaicin and chilli on haemorrhagic shock-induced gastric mucosal injury in the rat. *Journal of Gastroenterology and Hepatology* 13: 1007-1014.

Thomsen MK (1991). Substance P a neurogenic mediator of acute cellular inflammation in the dog? *Journal of Veterinary Pharmacology and Therapeutics* 14: 250–256.

Vikari A (2008). Positive results of plant extracts for beef cattle. *Feed-Mix* 16 (2): 34-35.

Watanabe T, Kawada T, Yamamoto M, Iwai K (1987). Capsaicin, a pungent principle of hot red pepper, evokes catecholamine secretion from the adrenal medulla of anesthetized rats. *Biochemical and Biophysical Research Communications* 142 (1): 259–264.

Weisshaar E, Heyer G, Forster C, Hardwerker HO (1998). Effect of topical caspaicin on the cutaneous reactions and itching to histamine in atopic eczema compared to healthy skin. *Archives of Dermatological Research* 290: 306–311.

Whiting S, Derbyshire E, Tiwari BK (2012). Capsaicinoids and capsinoids. A potential role for weight management? A systematic review of the evidence. *Appetite* 59 (2): 341-448.

Windisch W., Schedle K., Plitzner C., Kroismayer A (2008). Use of phytogenetic products as feed additives for swine and poultry. *Journal of Animal Science*, 86: E140–E148

Yoshimura M, Yonehara N, Ito T, Kawai Y, Tamura T (2000). Effects of topically applied capsaicin cream on neurogenic inflammation and thermal sensitivity in rats. *Japanese Journal of Pharmacology* 82: 116-121.

Yoshioka, M, St-Pierre S, Drapeau V, Dionne I, Doucet E, Suzuki M, Tremblay A (1999). Effects of red pepper on appetite and energy intake. *British Journal of Nutrition* 82 (2): 115-123.

Yu R, Park JW, Kurata T, Erickson KL (1998). Modulation of select immune responses by dietary capsaicin. *International Journal for Vitamin and Nutrition Research* 68: 114–119.

In: Capsaicin
Editor: Bruce Gilliam
ISBN: 978-1-61470-433-1

Chapter 3

VANILLOIDS AND THEIR EFFECT ON MAMMALIAN BIOLOGICAL BARRIERS

Mathias Kaiser and Francisco M. Goycoolea
Institute of Plant Biology and Biotechnology, Department of Biology,
Westfälische Wilhelms Universität Münster,
Laboratory of Nanobiotechnology, Münster, Germany

ABSTRACT

Capsaicinoids are a mixture of low molecular weight acid amides and vanillylamine and C_8 to C_{13} fatty acids that belong to the family of vanilloids. The major capsaicinoids naturally occurring in red chili peppers are capsaicin, dihydrocapsaicin and nordihydrocapsaicin, among other chain analogs that also occur in smaller amounts. Capsaicinoids have been reported to influence specific functions such as pain, body temperature regulation, body weight regulation, blood circulation and to exert a protective action to the integrity of the gastric mucosa. Recently, it has also been discovered that vanilloids are able to interfere with the cellular tight junctions (TJ) of several epithelial tissues. This makes these compounds potential candidates to enhance permeability of co-administered drugs across biological barriers like the blood-brain barrier or the intestinal barrier. Several *in vitro* studies have been carried out to investigate the interaction between capsaicin and the TJ of different cell lines. Complementary to permeability experiments, also electrical impedance measurements of cell monolayers and other assays were conducted. Based on these findings, the first attempts have been made to develop drug delivery systems which exploit the permeabilzing properties of the vanilloids. The scope of this review chapter is on an updated overview of the vanilloids which have been documented to interfere with cellular TJ, the proposed underlying mechanisms along with the main investigative methods and potential first applications of a recently discovered action of capsaicinoids.

1. INTRODUCTION

Capsaicin [I] and other vanilloids have been at the focus of attention due to the bioactivities that have been discovered besides their therapeutic use in chronic pain therapy. The potential of these substances in the treatment of different type of health disorders is yet to be fully realized. Recently, it has become evident that capsaicin induces the opening of epithelial TJ *in vitro*. The underlying mechanisms whereby capsaicin dislocates the proteins of the cytoskeleton have started to be elucidated. However, the implications of this effect in the permeabilization of poorly absorbable drugs with a therapeutic purpose are yet to be fully realized. Also, the inherent pungency of capsaicinoids precludes their use unless adequate drug delivery strategies are identified to overcome its burning sensation. This review aims to present an updated overview of the current state of knowledge regarding the effect of capsaicin in disrupting cellular TJ and the potential realization of this effect.

2. CAPSAICIN

Capsaicin and its analogues can be found in plants of the *capsicum* species [1], namely in chili peppers. Plants containing capsaicinoids have been used for centuries in traditional medicine like the Indian Ayurveda [2]. Being a lipophilic compound the solubility of capsaicin in water is very poor [3]. Capsaicin is able to desensitize the afferent sensory neurons. This phenomenon is called 'capsaicin desensitization' and was discovered by Szolcsányi in 1967 [4]. Later the responsible receptor was identified and cloned by Caterina *et al.*, [5]. This receptor is a non-selective cation channel which is structurally related to the TRP family. Initially called the vanilloid receptor 1 (VR1), today it is known as transient receptor potential vanilloid 1 (TRPV 1). This receptor is not only activated by capsaicin but also by acid and noxious heat [6], which explains the pungency of the compound. The pungency of capsaicin is measured by Scoville heat units (SHU) [7]. By Scoville's definition one ppm of capsaicin has 15 SHU (1). The most common medical applications of capsaicin are pain relieving formulations [8]. They can be used for the treatment of musculoskeletal pain but also for neuropathic pain [9]. For this purpose, topical applications like patches, gels or crèmes with different high doses of capsaicinoids can be applied. High doses of capsaicin can be used as long-term analgesia for neuropatic diseases if tolerated. Capsaicin also has an impact on body weight regulation [10]. It has been found out that activation of the TRPV 1 receptor prevents adipogenesis [11] and also increases the transcriptional and translational expression of glycolytic enzymes related to energy metabolism in human intestinal epithelial cells [12]. Furthermore, capsaicin has been used in therapy associated to its action on body temperature regulation [13], cancer treatment [14] urologic diseases [15] or also for mucus protection to prevent intestine inflammation [16]. Recently, it has also been found that capsaicin is able to reversibly open TJ [17]. This means that capsaicin is able to open a paracellular pathway into areas of the body which are protected by biological barriers. This fact makes it a very interesting molecule to increase the bioavailability of compounds.

At the same time, the inherent pungency and irritant action of these compounds in mammalians precludes their use in several applications. Hence, novel strategies aiming to modulate the pungency though not at the expense of losing the desired biological effect are needed. In capsicum fruits not only capsaicin but also its analogues can often occur [18]. The abundance of the different capsaicinoids is as follows: Capsaicin, 69%; dihydrocapsaicin [II], 22%; nordihydrocapsaicin [III], 7%; homocapsaicin [IV], 1%; and homodihydrocapsaicin [V], 1%. The first two are about twice as potent to the taste and nerves as the minor capsaicinoids.

Also there are synthetic derivatives of capsaicin. The most prominent compound is N-[(4-hydroxy-3-methoxyphenyl)methyl]-nonanamide [VI], also known as pelargonic acid vanillylamide, and is commonly named nonivamide. This compound has quite similar properties compared to capsaicin [19]. So it is also utilized in topical applications for pain relief for example [20, 21]. Based on nonivamide another substance called sodium nonivamide acetate (SNA) [VII] was developed [22]. In contrast to capsaicin it has no pungency as it does not increase the reaction latency in the rat model hot-plate and tail-flick assays. But interestingly SNA has anti-inflammatory properties that capsaicin does not show, like inhibition of vascular pain, hyperalgesia and edema of the hindpaw in rats. Also this compound is used for topical formulations [23]. There are also derivatives, which act as competitive antagonists of capsaicin.

The most famous compound is capsazepine [24] [VIII]. This synthetic analogue was able to reversibly reduce or abolish the current response to capsaicin of voltage-clamped dorsal root ganglion neurones from rats while the responses to γ-aminobutyric acid (GABA) and adenosine 5'-triphosphate (ATP) were unaffected. These results indicate that it is a specific antagonist of the actions of capsaicin sensory neuron. Another antagonist is olvanil [IX]. It is known that this analogue activates the TRPV1 receptor but with slower kinetics than capsaicin [6]. The reason for this behavior is probably its very high lipophilicity [25]. Capsiate [X] is yet another a compound of the family of capsinoids. It is a natural ingredient of the non-pungent pepper "CH-19 sweet" [25]. Instead of the amide bond between the vanillyl residue and the aliphatic chain it is liked by an ester bond. It activated the TRPV1 receptor of HEK293 cells in a similar manner as capsaicin did. Also induced nociceptive responses when injected into mice were the same. But when applied on skin, eyes or the oral cavity there was no response observable.

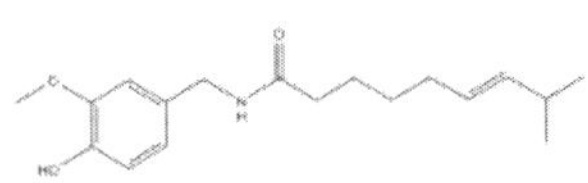

[I] Capsaicin

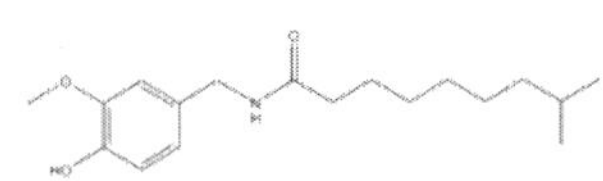

[II] Dihydrocapsaicin

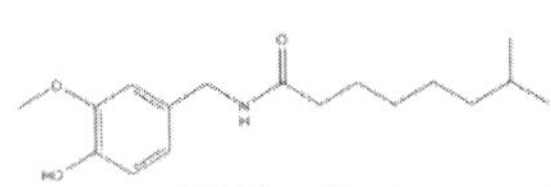

[III] Nordihydrocapsaicin

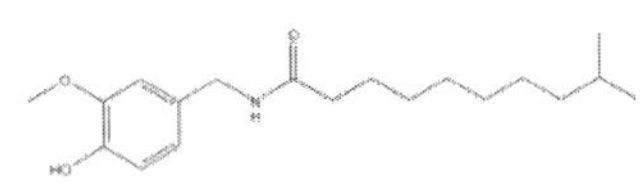

[IV] Homodihydrocapsaicin

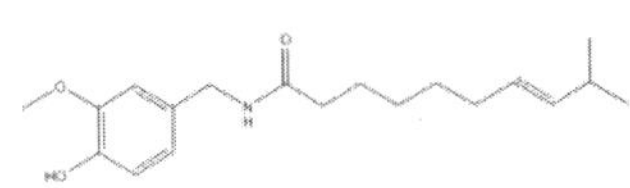

[V] Homocapsaicin

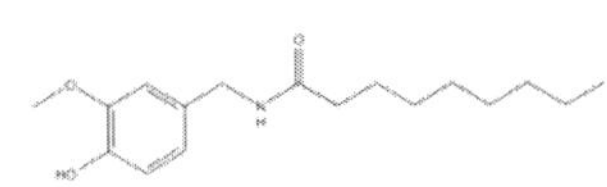

[VI] Nonivamide

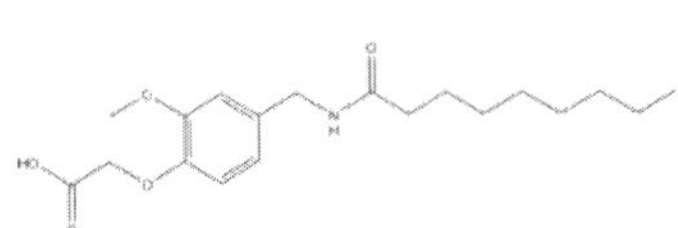

[VII] Nonivamide acetate

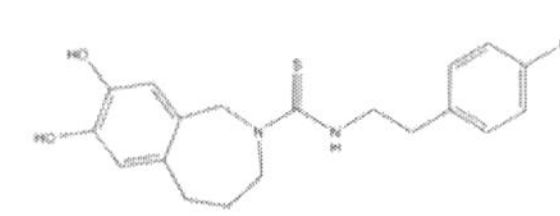

[VIII] Capsazepine

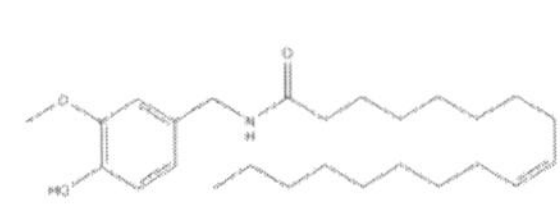

[IX] Olvanil

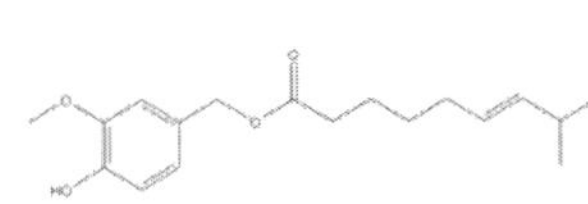

[X] Capsiate

Based on this evidence it was proposed that capsiate has to get direct access to the nerve endings to induce its effects. It was found that capsiate gets broken down during the experiments. After ~3 min the compound was completely cleaved in aqueous conditions. Stability seems to be a very important factor for the pungency of a substance.

3. Cell Tight Junctions

Tight junctions are expressed by most of the epithelial and endothelial cells [26]. They must be distinguished from adherens junctions which are mostly based on cadherin and catenin [27]. TJ are constructed by a complex network of different transmembrane proteins: Claudins [28], occludins [29] and junction adhesion molecules [30, 31] as the most important protein families. They are connected to intracellular proteins which belong to the zonula occludens (ZO) family [32, 33]. These link the whole construct to the actin skeleton of the cells [34] as depicted in Figure 1. Due to the establishment of this rigid macromolecular network there is no paracellular diffusion possible. This is the basis of the creation of an epithelial biological barrier. These barriers are key to protect the body against pathogens (bacteria, virus, fungal spores), foreign particles, large macromolecules and hydrophilic low molecular weight substances [35].

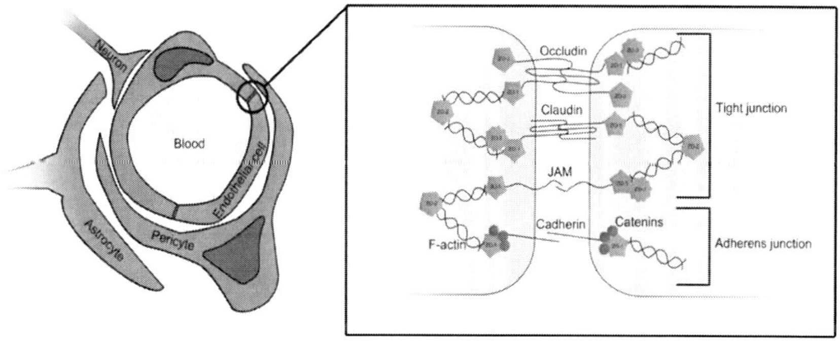

Figure 1. Schematic representation of a transverse section of a brain capillary showing the most important cell types involved in tight junction formation (left) and a simplified scheme showing the molecular composition of endothelial tight junctions and adherens junctions (right).

However, they also prevent the absorption of therapeutic drugs and hence reduce their bioavailability. This means that the only way for these compounds to overcome these epithelia is to exploit an active transport mechanism like carrier mediated transport or endocytosis [36]. Examples of barriers within the body that are built by a profuse 2D network of cell TJ include the blood brain barrier, the blood cerebrospinal fluid barrier, the bronchial airway epithelium and the intestinal barrier.

4. Methods to Investigate Integrity of Cell Tight Junctions

As capsaicin creates a paracellular pathway via opening of TJ, it is important to analyze the processes involved. The easiest way is to investigate the opening phenomenon itself. This can be done by simple white light microscopy or by transmission electron microscopy [37]. Recently, new techniques of light microscopy have been developed which can provide further information about the condition of the cells. One such technique is digital holographic microscopy (DHM) [38]. It is based on the classic holographic principle. It enables to do label-free, multi-focus quantitative phase imaging of living cells. Furthermore the cell volume and mass can be determined online after determination of the refractive index of the cells [39]. This provides information about a possible swelling or shrinking. For capsaicin it is known that due to calcium influx trough the TRPV 1 channel there is a swelling of the cells [5].

Another way of studying the effect of different substances, microorganisms or particles on cell TJ is to probe the integrity of confluent cell monolayers grown on artificial porous membranes. This membrane forms the floor of a fluid-filled chamber that is then inserted into a microplate well, such that the cell monolayer forms a barrier between the upper (luminal) and lower (basolateral) media chamber as shown on Figure 2 [40]. As model systems for biological barriers epithelial and endothelial cell lines are used. MDCK and Caco-2 cells are the most commonly used cell lines for this kind of assays as they form a very tight monolayer and are easy and rapid to cultivate [41, 42]. In the past these cell lines have been modified to form even stronger TJ like the MDCK-C7 clone [43]. Furthermore, they express influx and efflux transporters that make these systems also suitable for the study of active uptake mechanisms [44]. This setup allows several types of experiments. The mostly used one is the *in vitro* drug permeability assay. After placing the sample compound in the apical compartment samples can be taken from both chambers at several time points. From these aliquots the concentration of the compound can be determined. This allows the determination of the permeation coefficient P_{app} (cm/s) [45] as given below:

$$P_{app} = V_r \times \frac{dC}{dt} \times \frac{1}{AC_0}$$

where V_r (cm^3) is the volume of the acceptor compartment, *dC/dt* (M/s) is the slope of the cumulative concentration of the compound in the acceptor chamber over time, *A* (cm^2) is the membrane surface area and C_0 (M) is the initial concentration of the compound in the donor chamber [46].

Also inhibitors can be used to study the influence of different transporters or between active and passive transport [41]. This approach allows the use of specific inhibitors to study the influence of different transporters and whether the governing mechanism is mediated by active or passive transport. Moreover, when the compound of interest is applied at the basolateral side, it is possible to probe secretion processes. To probe the behavior of tight junctions this assay should be combined with the determination of the transepithelial (transendothelial) electrical resistance (TEER; $\Omega \cdot cm^2$).

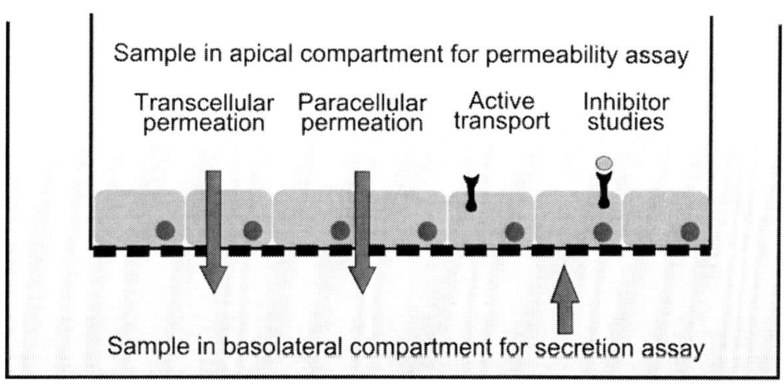

Figure 2. Side-view scheme of the set-up of an *in vitro* drug permeation experiment showing the donor (apical) and acceptor (basolateral) compartments separated by a cell monolayer overgrown on an artificial porous membrane. The scheme shows the different possible permeation mechanisms.

The first time TEER of brain microvascular endothelium was measured by Crone *in situ* using frogs as a model organism [47, 48]. In 1987 the first *in vitro* model with bovine brain microvascular cells was developed by Rutten [49]. Today the same filter membranes can be used for the TEER assay as well as the permeability assay. The most commonly used device is a voltohmeter fitted with two chop-sticks type electrodes. The electrodes are placed in the apical and basolateral chamber as shown in the inset in Figure 3 and a current is applied. Input resistance of the filter alone is subtracted from the values measured for the cell monolayers. The equivalent short-circuit current (I_{sc}) is calculated as TEEP/TEER, where TEEP is the transepithelial potential (mV).

Having lipophilic membranes the cell layer will hamper the flux of ions between the electrodes, which creates an electrical resistance that is recorded. A drug or substance that induces a change of the cell morphology which opens pathways for ions to cross the monolayer will result in a drop of electrical resistance. Recently, this technique has been improved further to allow automated measurements at programmed time intervals without need to remove the cells from the incubator nor disturb them during the measurements, which is known as impedance spectroscopy [50]. The experimental setup is the same as described above, but the frequency of the applied current is swept while measuring the amplitude and the phase of the resulting current. An example of an output impedance measurement is presented in Figure 3. The impedance obtained from this kind of measurement contains besides the TEER also information about the capacitance of the monolayer. This parameter can be used to further characterize the barrier properties of the monolayer. Another advantage of this technique is the possibility of performing continuous on-the-fly measurements, which is very important to probe reversible TJ opening effects.

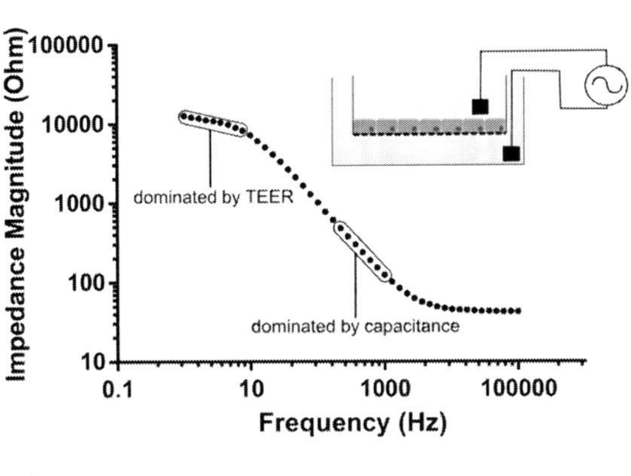

Figure 3. Representative example of the variation of the impedance with the frequency of the applied alternating current during measurement of a cell monolayer overgrown on an artificial porous membrane across two electrodes in a permeability chamber filled with medium. In the plot are shown the two regions of the curve from which the trans-epithelial/endothelial electrical resistance (TEER) and the capacitance are measured.

In addition to electrical methods, the tight junction can also be studied at the level of protein expression. Commonly used are electrophoresis assays to determine the concentration of the tight junction proteins [17]. Another frequently used method is immunofluorescence microscopy. This method was developed by Weber and his colleagues in the 1970s [51]. Specific fluorescent antibodies are used to stain the tight junction proteins as depicted in Figure 4. This method allows not only making conclusions about the amount of tight junction proteins but also to localize them within the cell.

5. Effects of Capsaicin on Cell Permeability *In Vitro*

The effect of capsaicin on cellular epithelium has been investigated for many cell lines. Isoda and colleagues studied extensively the behavior of Caco-2 cells. It was shown that the permeability of Caco-2 cells was increased significantly 45 min after treatment with 300 μM of capsaicin evidenced by TEER measurements [52]. This effect was not observed in presence of the competitive antagonist capsazepine. No remarkable changes in the intracellular Ca^{2+} concentration and LDH activity imply that this phenomenon is a reversible opening of the paracellular route and not due to any damage to the cell monolayer. In further studies they showed that the amount of F-actin is reduced which could be related to a reorganization of the cytoskeleton [53]. Also HSP47 was activated, which acts as a molecular chaperone during processing and/or secretion of procollagen. The ribosomal protein P2 also gets overexpressed after capsaicin treatment [54]. P2 activates elongation factor 2, which stabilizes F-actin filaments. This makes it is an important factor for the recovery of the TJ permeability. Yet another influencing factor is the dephosphorylation of cofilin [55]. It can directly initiate actin reorganization, which can open TJ. Also the group of Tanaka showed that capsaicin causes the opening of the intestinal epithelial TJ barrier at high doses [56]. Furthermore it inhibits P-glycoprotein function in the intestinal epithelium at low noncytotoxic doses. It is known to be involved in intestinal absorption and prehepatic elimination. So an inhibition could significantly enhance the bioavailability of drugs.

HCT-8 cells, a cell line from a human ileocoecal carcinoma, were used for the investigation of the impact of hot spices on the permeability of human intestinal epithelial monolayers [57]. It was shown that aqueous extracts of capsicum and cayenne pepper with a concentration of 7.5 mg/cm² significantly decreased the TEER and increased the permeability of 10-, 20- and 40-kDa dextrans but not those of 70 kDa. Extracts of other spices like black pepper, green pepper, bay leaf and nutmeg increased the TEER and did not have any effect on permeability.

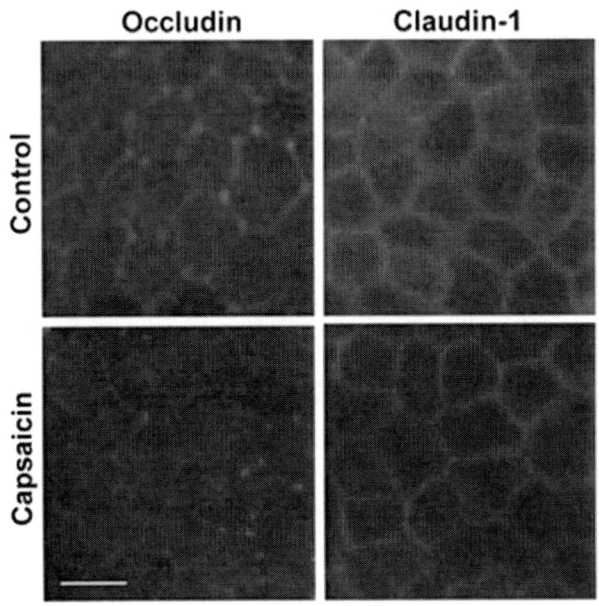

Figure 4. Confocal laser scanning microscope (CLSM) images of MDCK cells showing different immunostained tight junction proteins before (top images) and 45 minutes after (bottom images) treatment with capsaicin. Reproduced with permission from Shiobara *et al.*, Ref. 17.

Isoda's group in Japan also investigated the opening mechanism of TJ in MDCK cells [17]. Also here F-actin alteration was observed. The amount decreased in the bicellular junctions but was enhanced at the tricellular junctions and concentrated at the apical side of the lateral membrane. The location of the TJ proteins remained unchanged but the concentration of occludin was reduced significantly. Cofilin and LIM kinase are also involved. This mechanism of reversible TJ opening has not been observed for any other compound yet, which makes capsaicin a unique permeability enhancer. Capsaicin induces opening of tight junctions of MDCK cells when applied at sub-cytotoxic concentrations. This effect is reversible and after 360 minutes the TEER recovers to the original values. In our own studies, we have confirmed essentially the same effects using the same cell model as shown in Figure 5.

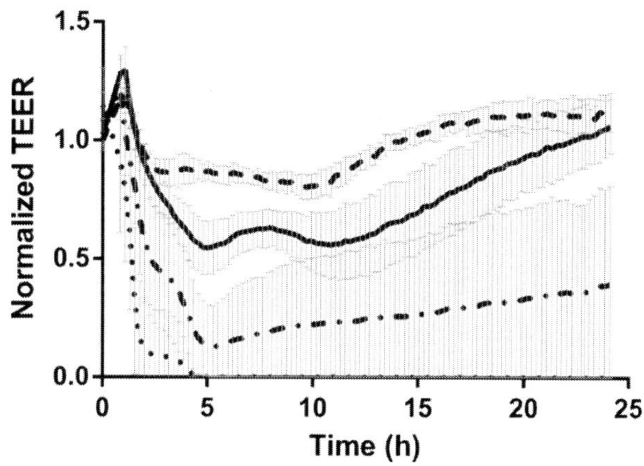

Figure 5. Relative variation with time of the TEER of MDCK-C7 cells after treatment with capsaicin at varying concentrations with respect to the untreated control. Applied concentrations of capsaicin were according with the following line code: Dashed – 250 mM; solid – 500 mM; dash-point – 750 mM; dotted – 1000 mM.

To date most of the therapeutic applications of capsaicin are based on its fundamental delivery. The permeability of capsaicin through various skin types has been evaluated using excised skin models derived from different organisms [58]. The study revealed that human skin is much less permeable for capsaicin that the skin of other animals like rat. Pig skin had quite comparable results and could be used as a model for skin penetration of capsaicin in humans. Interestingly, it was shown that the permeability of derivatives like nonivamide can vary drastically in relation to capsaicin. In another study the enhancing effect of capsaicin and nonivamide on the permeability of indomethacin, a non-steroidal anti-inflammatory drug, was evaluated [19]. Capsaicin and nonivamide significantly enhanced the flux of indomethacin across nude mouse skin in a concentration dependent manner. The enhancement of permeability was very similar for both compounds.

Permeation of capsaicin through cells of the oral cavity was also investigated. The submandibular gland (SMG) is an exocrine organ which secretes saliva and macromolecules under the regulation of the autonomic nerve system. It was shown by the group of Yu that primary SMG cells of rabbits express the TRPV 1 receptor [59]. Also they prove that capsaicin promoted salivary secretion. Later this group demonstrated that TRPV 1 is expressed in the human SMG as well [60]. TRPV1 activation directly upregulated expression of ZO-1, claudin-3, and -11, as well as increased paracellular permeability of TJs via myosin light chain 2 in normal and hyposecretory submandibular SMG cells in vitro [61]. For the rat SMG cell line SMG-C6 they demonstrated that in the presence of capsaicin the occludin fraction is redistributed while claudin-3, claudin-4 or E-cadherin were not affected [62]. Also here a redistribution of the actin skeleton and a concomitant decrease of the TEER were observed.

Furthermore, studies with other drugs show that stimulation of the TRPV family influences blood brain barrier (BBB) permeability [63]. In this experiment bEnd3, an immortalized mouse brain microvessel endothelial cell line and primary cells from mouse brain microvessels were used. Both cell types express the receptors TRPV2 and 4. The primary cells additionally express TRPV6. It was proposed that calcium influx plays a major role in BBB integrity.

6. Effects of Capsaicin on Permeability *In Vivo*

The response of central nervous system to capsaicin was investigated. In a study with rats capsaicin was applied to the sciatic nerve of rats [64]. The Evans blue assay, which was carried out 24 hours after capsaicin administration, showed a significantly higher concentration of dye in the spinal cord compared to the control. It was found that opening of the blood spinal cord barrier and the blood brain barrier follows activation of TRPV1-expressing C-fibers. The beginning of the opening is delayed by many hours after the administration and remains for several days. Also another study confirmed that if capsaicin is applied to the brain surface the permeability of the pial microvasculature increased [65]. It also demonstrated that a neurokinin 1 agonist, probably substance P, is a mediator for that process. Interestingly it was also shown that an intraplantar administration of capsaicin in rats did not induce changes of occludin in lumbar spinal cord [66].

Also the influence of capsaicin on the intestinal drug absorption process was valuated in the rat [67]. It was shown that in presence of capsaicin the uptake of the antibiotic cephalexin was reduced. It is known that this drug is taken up by the H^+/peptide co-transporter 1 (PEPT1), which is dependent on Ca^{2+} concentration [68]. But it was also revealed that the permeability of cephalexin via the paracellular pathway in the rat jejunum is increased by capsaicin [67].

The effect of capsaicin on airway sensory nerves in the human nose was studied [69]. For this study four healthy and four volunteers with perennial allergic rhinitis received capsaicin nasal challenges with a concentration of 20 μmol/L. Capsaicin did not increase vascular permeability in the airway, because the albumin content of nasal lavage fluid was not increased. This finding is at odds with animal studies where an increase of vascular permeability was observed [70, 71].

7. Capsaicin in Drug Delivery Systems for Improvement of Bioavailability of Drugs

Several nano- and microparticulate systems containing capsaicin have been developed in recent years [72-79]. Most of these formulations are based on natural polymers and have the purpose to make capsaicin compatible to the aqueous environment in organisms due to its low water solubility [3]. The use of capsaicin loaded drug delivery systems to improve the bioavailability of other drugs has not been extensively studied so far. The group of Shin developed a capsaicin containing nanoemulsion based on oleoresin of capsicum which contains 22.67 mg/g of capsaicin and the nonionic surfactants Tween 80 and Span 80 [79]. The mean drop size ranged from 20 to 62 nm. Skin permeation assays revealed that the formulation was able to permeate all skin layers from the stratum corneum to the dermis. The group suggested that the developed nano system is a potential candidate for the transdermal delivery of drugs. The group of Gwak tested capsaicin loaded nano formulations based on alginate and chitosan *in vivo* with rats and evaluated their pharmacokinetics (PK) [78]. Coated nanoemulsions with alginate, chitosan or a mix of both polymers were prepared. The size of the formulations ranged between ~50 and ~100 nm and zeta potential either positive (~ + 36 mV) or negative (~ - 42 mV) when the polymer at the surface was chitosan or alginate, respectively. Significant differences in comparison to the control group were observed. Though in general, the relative bioavailability of formulated nanoemulsions was up to 131.7 times higher, differences in the PK profiles of capsaicin associated to the composition of these nanosystems were reported. According with the PK results there were not significant differences in the major parameters between alginate- and chitosan-based systems. However, in comparison to the double layer nano-emulsions the triple layer nanoemulsions showed an overall lower bioavailability. In our group capsaicin loaded core shell nanocapsules with an oily core and a chitosan coating were developed. The size of the produced capsules ranged in size between ~160 and ~250 nm and had a zeta potential of ~ + 65 mV. In TEER measurements using impedance spectroscopy with MDCK cells there was a reversible TJ opening effect visible. Capsaicin remains active also in in the nanosystem and encapsulation modifies the TEER behavior. This effect was confirmed with DHM microscopy.

CONCLUSION

Capsaicin is a natural therapeutic molecule with a large scope of therapeutic potential which is even expanded by its chemical analogues. However up to now its use as an enhancer of drug bioavailability has only started to be realized. Only few drug delivery formulations have so far been developed that exploit the capability of capsaicin to open TJ and hence favor the paracellular transport. In the future the feasibility of these systems has to be evaluated, especially the route of administration plays an important role. If new systems are able to increase the bioavailability of drug using a very convenient way of application they could revolutionize the design of commercial drugs.

ACKNOWLEDGMENTS

We acknowledge funding support from the DFG (Project GRK 1549 IRTG 'Molecular and Cellular GlycoSciences') and the Danish Agency for Science, Technology and Innovation (FENAMI project 10-093456).

REFERENCES

[1] Wahyuni Y, Ballester A-, Sudarmonowati E, Bino RJ, Bovy AG. Secondary metabolites of capsicum species and their importance in the human diet. *J. Nat. Prod.* 2013;76(4):783-93.

[2] Meghvansi MK, Siddiqui S, Khan MH, Gupta VK, Vairale MG, Gogoi HK, *et al.*, Naga chilli: A potential source of capsaicinoids with broad-spectrum ethnopharmacological applications. *J. Ethnopharmacol.* 2010 OCT 28;132(1):1-14.

[3] Turgut C, Newby B-, Cutright TJ. Determination of optimal water solubility of capsaicin for its usage as a non-toxic antifoulant. *Environmental Science and Pollution Research*. 2004;11(1):7-10.

[4] Jancsó N, Jancsó-Gábor A, Szolcsányi J. Direct evidence for neurogenic inflammation and its prevention by denervation and by pretreatment with capsaicin. *Br. J. Pharmacol.* 1967;31(1):138-51.

[5] Caterina MJ, Schumacher MA, Tominaga M, Rosen TA, Levine JD, Julius D. The capsaicin receptor: A heat-activated ion channel in the pain pathway. *Nature.* 1997;389(6653):816-24.

[6] Julius D, Basbaum AI. Molecular mechanisms of nociception. *Nature.* 2001;413(6852):203-10.

[7] Scoville WL. Note on capsicums. *J. Am. Pharm. Assoc.* 1912;1(5):453-4.

[8] Smith H, Brooks JR. Capsaicin-based therapies for pain control. In: Abdel-Salam OME, editor. *Capsaicin as a Therapeutic Molecule.* Springer Basel; 2014. p. 129-46.

[9] Schumacher M, Pasvankas G. Topical capsaicin formulations in the management of neuropathic pain. In: Abdel-Salam OME, editor. *Capsaicin as a Therapeutic Molecule*. Springer Basel; 2014. p. 105-28.

[10] Leung FW. Capsaicin as an anti-obesity drug. In: Abdel-Salam OME, editor. *Capsaicin as a Therapeutic Molecule*. Springer Basel; 2014. p. 171-9.

[11] Zhang LL, Liu DY, Ma LQ, Luo ZD, Cao TB, Zhong J, *et al.*, Activation of transient receptor potential vanilloid type-1 channel prevents adipogenesis and obesity. *Circ. Res.* 2007;100(7):1063-70.

[12] Han J, Isoda H. Capsaicin induced the upregulation of transcriptional and translational expression of glycolytic enzymes related to energy metabolism in human intestinal epithelial cells. *J. Agric. Food Chem.* 2009;57(23):11148-53.

[13] Steiner AA, Turek VF, Almeida MC, Burmeister JJ, Oliveira DL, Roberts JL, *et al.*, Nonthermal activation of transient receptor potential vanilloid-1 channels in abdominal viscera tonically inhibits autonomic cold-defense effectors. *Journal of Neuroscience*. 2007;27(28):7459-68.

[14] Díaz-Laviada I, Rodríguez-Henche N. The potential antitumor effects of capsaicin. In: Abdel-Salam OME, editor. *Capsaicin as a Therapeutic Molecule*. Springer Basel; 2014. p. 181-208.

[15] Foster Jr. HE, Lake AG. Use of vanilloids in urologic disorders. In: Abdel-Salam OME, editor. *Capsaicin as a Therapeutic Molecule*. Springer Basel; 2014. p. 307-17.

[16] Mózsik G, Past T, Habon T, Keszthelyi Z, Perjési P, Kuzma M, *et al.*, Capsaicin is a new gastrointestinal mucosal protecting drug candidate in humans. In: Gyula Mozsik, editor. *Capsaicin - Sensitive Neural Afferentation and the Gastrointestinal Tract: from Bench to Bedside*. AvE4EvA; 2014. p. 209-308.

[17] Shiobara T, Usui T, Han J, Isoda H, Nagumo Y. The reversible increase in tight junction permeability induced by capsaicin is mediated via cofilin-actin cytoskeletal dynamics and decreased level of occludin. *PLoS ONE*. 2013;8(11).

[18] Cordell GA, Araujo OE, Burgos GAS, Perreault M. Capsaicin: Identification, nomenclature, and pharmacotherapy. *Ann. Pharmacother.* 1993;27(3):330-6.

[19] Fang J-, Fang C-, Hong C-, Chen H-, Lin T-, Wei H-. Capsaicin and nonivamide as novel skin permeation enhancers for indomethacin. *European Journal of Pharmaceutical Sciences*. 2000;12(3):195-203.

[20] Lunter D, Daniels R. In vitro skin permeation and penetration of nonivamide from novel film-forming emulsions. *Skin Pharmacology and Physiology*. 2013;26(3):139-46.

[21] Rottke M, Lunter DJ, Daniels R. In vitro studies on release and skin permeation of nonivamide from novel oil-in-oil-emulsions. *European Journal of Pharmaceutics and Biopharmaceutics*. 2014;86(2):260-6.

[22] Yang J-, Tang C-, Yu K-, Yang Y-, Chang W-, Chen I-. Sodium nonivamide acetate: A non-pungently antinociceptive capsaicin derivative with unusual anti-inflammatory properties. *Gen Pharmacol.* 1996;27(1):141-50.

[23] Huang YB, Lin YH, Lu TM, Wang RJ, Tsai YH, Wu PC. Transdermal delivery of capsaicin derivative-sodium nonivamide acetate using microemulsions as vehicles. *Int. J. Pharm*. 2008;349(1-2):206-11.

[24] Bevan S, Hothi S, Hughes G, James IF, Rang HP, Shah K, *et al.*, Capsazepine: A competitive antagonist of the sensory neurone excitant capsaicin. *Br. J. Pharmacol*. 1992;107(2):544-52.

[25] Iida T, Moriyama T, Kobata K, Morita A, Murayama N, Hashizume S, *et al.*, TRPV1 activation and induction of nociceptive response by a non-

pungent capsaicin-like compound, capsiate. *Neuropharmacology*. 2003 6;44(7):958-67.

[26] Van Itallie CM, Anderson JM. Claudins and epithelial paracellular transport. *Annual Review of Physiology*. 2006;68:403-29.

[27] Yap AS, Brieher WM, Gumbiner BM. Molecular and functional analysis of cadherin-based adherens junctions. *Annual Review of Cell and Developmental Biology*. 1997;13:119-46.

[28] Furuse M, Fujita K, Hiiragi T, Fujimoto K, Tsukita S. Claudin-1 and -2: Novel integral membrane proteins localizing at tight junctions with no sequence similarity to occludin. *J. Cell Biol*. 1998;141(7):1539-50.

[29] Furuse M, Hirase T, Itoh M, Nagafuchi A, Yonemura S, Tsukita S, *et al*., Occludin: A novel integral membrane protein localizing at tight junctions. *J Cell Biol*. 1993;123(6 PART 2):1777-88.

[30] Martìn-Padura I, Lostaglio S, Schneemann M, Williams L, Romano M, Fruscella P, *et al*., Junctional adhesion molecule, a novel member of the immunoglobulin superfamily that distributes at intercellular junctions and modulates monocyte transmigration. *J. Cell Biol.* 1998;142(1):117-27.

[31] Aurrand-Lions M, Duncan L, Ballestrem C, Imhof BA. JAM-2, a novel immunoglobulin superfamily molecule, expressed by endothelial and lymphatic cells. *J. Biol. Chem*. 2001;276(4):2733-41.

[32] Stevenson BR, Siliciano JD, Mooseker MS, Goodenough DA. Identification of ZO-1: A high molecular weight polypeptide associated with the tight junction (zonula occludens) in a variety of epithelia. *J. Cell Biol.* 1986;103(3):755-66.

[33] Gumbiner B, Lowenkopf T, Apatira D. Identification of a 160-kDa polypeptide that binds to the tight junction protein ZO-1. *Proc. Natl. Acad. Sci. USA*. 1991;88(8):3460-4.

[34] Tsukita S, Furuse M, Itoh M. Multifunctional strands in tight junctions. Nature Reviews Molecular *Cell Biology*. 2001;2(4):285-93.

[35] Sonaje K, Lin K, Wang J, Mi F, Chen C, Juang J, *et al*., Self-assembled pH-sensitive nanoparticles: A platform for oral delivery of protein drugs. *Advanced Functional Materials*. 2010;20(21):3695-700.

[36] Chen Y, Liu L. Modern methods for delivery of drugs across the blood–brain barrier. *Adv. Drug Deliv. Rev*. 2012 5/15;64(7):640-65.

[37] Yeh T, Hsu L, Tseng MT, Lee P, Sonjae K, Ho Y, *et al*., Mechanism and consequence of chitosan-mediated reversible epithelial tight junction opening. *Biomaterials*. 2011 9;32(26):6164-73.

[38] Kemper B, Carl D, Höink A, Von Bally G, Bredebusch I, Schnekenburger J. Modular digital holographic microscopy system for marker free quantitative phase contrast imaging of living cells. Progress in biomedical optics and imaging - proceedings of SPIE; ; 2006.

[39] Barer R. Interference microscopy and mass determination. *Nature.* 1952;169(4296):366-7.

[40] Siflinger-Birnboim A, Del Vecchio PJ, Cooper JA, Blumenstock FA, Shepard JM, Malik AB. Molecular sieving characteristics of the cultured endothelial monolayer. *J. Cell Physiol.* 1987;132(1):111-7.

[41] Volpe DA. Drug-permeability and transporter assays in caco-2 and MDCK cell lines. *Future Medicinal Chemistry.* 2011;3(16):2063-77.

[42] Irvine JD, Takahashi L, Lockhart K, Cheong J, Tolan JW, Selick HE, *et al.*, MDCK (madin-darby canine kidney) cells: A tool for membrane permeability screening. *J. Pharm Sci.* 1999;88(1):28-33.

[43] Gekle M, Wünsch S, Oberleithner H, Silbernagl S. Characterization of two MDCK-cell subtypes as a model system to study principal cell and intercalated cell properties. *Pflügers Archiv European Journal of Physiology.* 1994;428(2):157-62.

[44] Volpe DA. Variability in caco-2 and MDCK cell-based intestinal permeability assays. *J. Pharm Sci.* 2008;97(2):712-25.

[45] van Breemen R,B., Li Y. Caco-2 cell permeability assays to measure drug absorption. *Expert Opin. Drug Metab. Toxicol.* 2005 08/01; 2014/07;1(2):175-85.

[46] Artursson P, Karlsson J. Correlation between oral drug absorption in humans and apparent drug permeability coefficients in human intestinal epithelial (caco-2) cells. *Biochem. Biophys. Res. Commun.* 1991 3/29;175(3):880-5.

[47] Crone C, Christensen O. Electrical resistance of a capillary endothelium. *The Journal of General Physiology.* 1981 April 01;77(4):349-71.

[48] Crone C, Olesen SP. Electrical resistance of brain microvascular endothelium. *Brain Res.* 1982 6/3;241(1):49-55.

[49] Rutten MJ, Hoover RL, Karnovsky MJ. Electrical resistance and macromolecular permeability of brain endothelial monolayer cultures. *Brain Res.* 1987;425(2):301-10.

[50] Benson K, Cramer S, Galla H-. Impedance-based cell monitoring: Barrier properties and beyond. *Fluids and Barriers of the CNS.* 2013;10(1).

[51] Lazarides E, Weber K. Actin antibody: The specific visualization of actin filaments in non-muscle cells. *Proceedings of the National Academy of Sciences*. 1974 June 01;71(6):2268-72.

[52] Isoda H, Han J, Tominaga M, Maekawa T. Effects of capsaicin on human intestinal cell line caco-2. *Cytotechnology*. 2001;36(1-3):155-61.

[53] Han J, Isoda H, Maekawa T. Analysis of the mechanism of the tight-junctional permeability increase by capsaicin treatment on the intestinal caco-2 cells. *Cytotechnology*. 2003;40(1-3):93-8.

[54] Han J, Akutsu M, Talorete TPN, Maekawa T, Tanaka T, Isoda H. Erratum: Capsaicin-enhanced ribosomal protein P2 expression in human intestinal caco-2 cells (cytotechnology 47 (89-96)). *Cytotechnology*. 2005;49(1):77-85.

[55] Nagumo Y, Han J, Arimoto M, Isoda H, Tanaka T. Capsaicin induces cofilin dephosphorylation in human intestinal cells: The triggering role of cofilin in tight-junction signaling. *Biochem. Biophys. Res. Commun.* 2007 4/6;355(2):520-5.

[56] Tsukura Y, Mori M, Hirotani Y, Ikeda K, Amano F, Kato R, *et al.*, Effects of capsaicin on cellular damage and monolayer permeability in human intestinal caco-2 cells. *Biological and Pharmaceutical Bulletin.* 2007;30(10):1982-6.

[57] Jensen-Jarolim E, Gajdzik L, Haberl I, Kraft D, Scheiner O, Graf J. Hot spices influence permeability of human intestinal epithelial monolayers. *J. Nutr*. 1998;128(3):577-81.

[58] Fang J-, Wu P-, Huang Y-, Tsai Y-. In vitro permeation study of capsaicin and its synthetic derivatives from ointment bases using various skin types. *Int. J. Pharm*. 1995;126(1-2):119-28.

[59] Zhang Y, Xiang B, Li Y, Wang Y, Wang X, Wang Y, *et al.*, Expression and characteristics of vanilloid receptor 1 in the rabbit submandibular gland. *Biochem. Biophys. Res. Commun*. 2006 6/23;345(1):467-73.

[60] Ding QW, Zhang Y, Wang Y, Wang YN, Zhang L, Ding C, *et al.*, Functional vanilloid receptor-1 in human submandibular glands. *J. Dent Res*. 2010;89(7):711-6.

[61] Cong X, Zhang Y, Shi L, Yang N-, Ding C, Li J, *et al.*, Activation of transient receptor potential vanilloid subtype 1 increases expression and permeability of tight junction in normal and hyposecretory submandibular gland. *Laboratory Investigation*. 2012;92(5):753-68.

[62] Cong X, Zhang Y, Yang N-, Li J, Ding C, Ding Q-, *et al.*, Occludin is required for TRPV1-modulated paracellular permeability in the submandibular gland. *J. Cell Sci.* 2013;126(5):1109-21.

[63] Brown RC, Wu L, Hicks K, O'Neil RG. Regulation of blood-brain barrier permeability by transient receptor potential type C and type V calcium-permeable channels. *Microcirculation.* 2008;15(4):359-71.

[64] Beggs S, Liu XJ, Kwan C, Salter MW. Peripheral nerve injury and TRPV1-expressing primary afferent C-fibers cause opening of the blood-brain barrier. *Molecular Pain.* 2010;6.

[65] Hu D, Easton A, Fraser P. TRPV1 activation results in disruption of the blood-brain barrier in the rat. *Br. J. Pharmacol.* 2005 OCT;146(4):576-84.

[66] Xanthos DN, Püngel I, Wunderbaldinger G, Sandkühler J. Effects of peripheral inflammation on the blood-spinal cord barrier. *Molecular Pain.* 2012;8.

[67] Komori Y, Aiba T, Sugiyama R, Nakai C, Kawasaki H, Kurosaki Y. Effects of capsaicin on intestinal cephalexin absorption in rats. *Biological and Pharmaceutical Bulletin.* 2007;30(3):547-51.

[68] Wenzel U, Kuntz S, Diestel S, Daniel H. PEPT1-mediated cefixime uptake into human intestinal epithelial cells is increased by Ca2+ channel blockers. *Antimicrob Agents Chemother.* 2002;46(5):1375-80.

[69] Philip G, Baroody FM, Proud D, Naclerio RM, Togias AG. The human nasal response to capsaicin. *J. Allergy Clin. Immunol.* 1994;94(6 I):1035-45.

[70] Asakura K, Shirasaki H, Narita S, Kojima T, Kataura A. Study on the dye leakage response of nasal mucosa following topical, capsaicin challenge in guinea pigs. *Acta Otolaryngol.* 1992;112(3):545-51.

[71] Lundblad L, Saria A, Lundberg JM, Anggard A. Increased vascular permeability in rat nasal mucosa induced by substance P and stimulation of capsaicin-sensitive trigeminal neurons. *Acta Otolaryngol.* 1983;96(5-6):479-84.

[72] Xing F, Cheng G, Yang B, Ma L. Microencapsulation of capsaicin by the complex coacervation of gelatin, acacia and tannins. *J. Appl. Polym. Sci.* 2004;91(4):2669-75.

[73] Xing F, Cheng G, Yi K. Study on the antimicrobial activities of the capsaicin microcapsules. *J. Appl. Polym. Sci.* 2006;102(2):1318-21.

[74] Choi A-, Kim C-, Cho Y-, Hwang J-, Kim C-. Characterization of capsaicin-loaded nanoemulsions stabilized with alginate and chitosan by self-assembly. *Food and Bioprocess Technology.* 2011;4(6):1119-26.

[75] Goycoolea FM, Valle-Gallego A, Stefani R, Menchicchi B, David L, Rochas C, *et al.*, Chitosan-based nanocapsules: Physical

characterization, stability in biological media and capsaicin encapsulation. *Colloid Polym. Sci.* 2012;290(14):1423-34.

[76] Jincheng W, Sihao C. Preparation and characterization of microcapsules containing capsaicin. *J. Appl. Polym. Sci.* 2010;116(4):2234-41.

[77] Zhu Y, Peng W, Zhang J, Wang M, Firempong CK, Feng C, *et al.*, Enhanced oral bioavailability of capsaicin in mixed polymeric micelles: Preparation, in vitro and in vivo evaluation. *Journal of Functional Foods.* 2014;8(1):358-66.

[78] Choi AY, Kim C-, Park HY, Kim HO, Lee NR, Lee KE, *et al.*, Pharmacokinetic characteristics of capsaicin-loaded nanoemulsions fabricated with alginate and chitosan. *J. Agric. Food Chem.* 2013;61(9):2096-102.

[79] Kim JH, Ko JA, Kim JT, Cha DS, Cho JH, Park HJ, *et al.*, Preparation of a capsaicin-loaded nanoemulsion for improving skin penetration. *J. Agric. Food Chem.* 2014;62(3):725-32.

In: Capsaicin
Editor: Bruce Gilliam

ISBN: 978-1-61470-433-1

Chapter 4

ANALGESIC EFFECTS OF *CAPSICUM FRUTESCENS* LINN. [SOLANACEAE] FRUIT AQUEOUS EXTRACT IN MICE

A. T. Jolayemi[*1] *and J. A. O. Ojewole* [2]
[1]Department of Anaesthesia and ICU, Goulburn Base Hospital and Senior Lecturer Australia National University (ANU), Goulburn City, NSW, Australia
[2]Department of Pharmacology Faculty of Health Sciences, University of KwaZulu Natal, Durban, South Africa

ABSTRACT

The analgesic property of Capsicum frutescens LINN. SOLANACEAE] extract-CFE (98% Capsaicin by NMR) of South African origin in mice (20-30 g), using 'hot plate' and 'acetic acid' methods was investigated. In the hot plate test, twelve mice in each of 5 groups were administered intraperitoneal (i.p.) CFE or synthetic capsaicin (Fluka Biotechnika-CF) at 0.5-8 mg/kg and at hourly intervals; at 40 C, the mean reaction time (MRT) was estimated. In the 'acetic acid' test method, CFE or CF was administered imp. at various doses (0.5- 8 mg/Kg) at hourly intervals following single treatment using 0.2 mls of 3

* Corresponding author: Dr Adebayo Taiwo Ezekiel Jolayemi E-mail: adebayoj@hotmail.com Department of Anaesthesia and ICU Goulburn Hospital NSW, Senior Lecturer, Australia National University.

% v/v acetic acid-induced writhing responses, or non-treated (n=12). Separate groups of rats (n=12) were pre-treated with 100-mg/kg diclofenac (DIC) or morphine (MPN) 10 mg/kg i.p. to evaluate the peripheral analgesic effects and central analgesia, respectively. Data obtained were pooled and analysed by the student's T-test. The 'hot plate' and 'acetic acid' test methods showed significant ($p<0.0001$) prolongation of the MRT and inhibition of writhing responses, respectively, compared to the control. Similar results with a Pearson correlation of 0.999 were obtained in the CF subgroups. This study shows that capsaicin has superior central analgesia compared to equipotency doses of MPN or DIC and comparable peripheral analgesia to either MPN or DIC.

1.1. Introduction

Chronic pain is a major health hazard resulting in debilitation, sickness role, and adverse drug reactions, including death (Jolayemi, 2002). Pain management is multidisciplinary, with current trend emphasizing complimentary health methods (Wallace, 1997). This multi-specialty approach is unaffordable for patients in developing countries (due to poor socio-economic conditions), resulting in poor compliance, self-medication and high complication rate, or death. A review of the evolution of analgesics shows that they are derived primarily from basic foods, natruceaticals, and other edible or chewable substances, and later from conventional medications. For example, eating poppy leaf and cannabinoid tea, or chewing of opioids or addictive substances, have been known to certain racial groups for centuries. It is, therefore, important to use these natural interventions in a more scientific and beneficial way to improve compliance and health; and reduce the unwanted side-effects of conventional analgesics. The development of inhibitors of Substance P of plant origin (reviewed by Bevan et al., 1995) provides an opportunity in the treatment of intractable pain.

Capsicum species occur worldwide, and has been used for more than 9000 years by the Chinese, Indians, and Africans for medicinal and non-medicinal purposes (Watt, 1962). One of the main objectives of this study was to determine if *Capsicum frutescens* fruit extract of South African origin, has similar efficacy on peripheral and central components of pain as described for *Capsicum spp* (Linn) [Solanaceae] in other parts of the world, such as India, Mexico, Thailand and South America (Jaiarj et al., 1998).

Studies described in the literature suggest that capsaicin could modulate analgesia (Bevan, 1990; Winter, 1997). Capsaicin extracted from *Capsicum spp* act at the vanilloid receptors to inhibit Substance P, and has proved very useful in intractable pain of diabetic neuropathy, as well as herpetic and trigeminal neuralgia (Davis, 2000). Most of these studies have shown the analgesic effect of capsaicin in spinal mechanism. The questions that remain unanswered include elucidation of central and peripheral analgesic mechanisms of capsaicin. The protection of pre-treated rats from thermal pain in the 'hot plate' test method as described by (Ojewole, 2002) and the percentage inhibition of writhing in the pre-treated rats using the 'acetic acid' test method were used in this study.

1.2. Animals

Experiments were performed on mice (20-30 g body weight) of both sexes. The mice were maintained under standard laboratory conditions of light, temperature and relative humidity. The animals were fed with standard diet pellet (Epol-diet 4700, Epol, South Africa) and drinking tap water *ad libitum*. They were kept and maintained in Biomedical Resource Unit, University of KwaZulu-Natal, Durban 4000, South Africa.

1.2.1. Drugs

Investigative drugs included capsaicin (Fluka-Biotechnika), *Capsicum frutescens*-derived extract, and positive controls (morphine) and diclofenac.

1.2.2. Assessment of Analgesic Effects of Capsaicin Using the 'Hot Plate' Test Method

The 'hot-plate' (thermal) test method employed in this study was modified from those described in detail earlier by Eddy and Leimback, (1953) cited by Ojewole, (2006); Lanhers et al., (1992); and Williamson et al., (1996) cited by Ojewole, (2006). Mice weighing 20–30 g were divided into two groups (A and B) of control and treated animals. Group A had three subgroups of 12 mice each. The first subgroup consisted of untreated mice, the mice in second and third subgroups were treated with morphine (MPN, 10 mg/kg i.p.) and diclofenac, (DIC 100 mg/kg i.p.), respectively. Group B consisted of 2 main subgroups treated with *Capsicum* fruit aqueous extract (CFE) and CPF as the reference drug. Each subgroup was subdivided into 5 subunits based on the dose regimen of 0.5-8 mg/kg i.p. The animals had similar demographics in terms of sex, age and weight. All the animals were fed with standard diet pellet and drinking tap water *ad libitum*. A 600-ml glass beaker was placed on a 'Heidolph® MR 2002' hot-plate (with adjustable temperature). The temperature of the hot-plate was then regulated to $40 \pm 1^{o}C$. Each mouse was placed in the glass beaker (on the hot-plate) in order to obtain the animal's response to electrical heat-induced nociceptive pain (licking of the forepaws and eventually jumping out of the glass beaker). Jumping out of the beaker was taken as an indicator of the animal's response to heat-induced nociceptive pain. The time taken for each mouse to jump out of the beaker (i. e. reaction time) was noted and recorded. Capsaicin extract and CPF were tested at doses of 0.50, 1.0, 2.0, 4, and 8.0 mg/kg i. p., respectively. Treated 'control' mice received MPN (10 mg/kg i. p.), or DIC (100 mg/kg route) only. Due to the effects of gastric juice, capsaicin was administered by intraperitoneal route, a route that ensures adequate systemic bioavailability, early onset and satisfactory pharmacological activity of the tested agents.

The mean reaction time for each 'test' mouse was determined and documented using a standard stop watch calibrated in seconds. Each mouse served as its own control. Thus, before treatment, its reaction time was determined thrice at 1-hour intervals. The mean of these three determinations constituted the 'initial reaction time'– that is, reaction time before the treatment of the mouse. The mean reaction times for all the mice used were pooled to obtain the final, 'control' mean reaction time (t_1). Each of the mice in the study was thereafter treated with either CFE or CPF. Thirty minutes after treatment with CFE or CPF, the reaction time was again evaluated (t_2). This value was pooled for the mice used in each treatment group, and the final 'test' mean value for each treatment group was calculated. This final 'test' mean value represented 'after-treatment reaction time' for each group of treated mice. This 'test' mean value was subsequently used to determine the percentage thermal pain stimulus relief or protection, by applying the formula:

$$\%\text{ protection against thermal pain stimulus} = \text{Test Mean} - \text{Control Mean X } 100/\text{Control Mean}$$

$$= t_1 - t_2 \text{ x } 100 / t_2$$

Where t_1 = the mean 'control' reaction time, and t_2 = mean 'test' reaction time, respectively.

1.2.3. Assessment of CFE and CPF Analgesia Using 'Acetic Acid' Test Method

The 'acetic acid' test method used in this study was adopted from those described earlier by Koster et al. (1959); Williamson et al. (1996); Zakaria et al. (2001); and Silva et al. (2003). The mice used were divided into groups of 'test' and 'control' animals. The treated group had subunits that received CFE and CPF, respectively, subdivided into 5 treatment groups of 10 mice each for graded doses of CFE or CPF at 0.5, 1, 2, 4 and 8.0 mg/kg (i.p). Group A control mice included an untreated group of 12 mice and two positive 'control' groups of 10 mice each, that were pre-treated with DIC (100 mg/kg i.p.) and MPN (10 mg/kg i.p.). Each of the untreated 'control', and 'test' animals was treated with intraperitoneally-administered 0.2 ml of 3% *v*/v acetic acid solution (Koster et al. 1959). Twenty minutes after pre-treatment with DIC, a dose of the capsaicin, or the extract, 0.2 ml of a 3% w/v of acetic acid solution was injected intraperitoneally (i. p.) to each of the 'treated' mice (Koster et al. 1959). The number of writhings (i. e., abdominal contractions and stretches) that occurred within the first 20 minutes following acetic acid administration were counted and recorded. The recorded numbers of acetic acid-induced writhings (abdominal contractions and stretches) that occurred in the CFE-, CPF-, and DIC- pretreated mice were compared with those in the 'untreated Group A 'control' mice. This 'test' mean value was subsequently used to determine percentage inhibition to writhing response using the formula:

% inhibition of writhings response= Control Mean – Test Mean X 100/Control Mean

$$= w_1 - w_2 \text{ x } 100 / w_1$$

Where w_1 = mean 'control' and w_2 = mean 'test' writhing movements, respectively.

1.2.4. Data Analysis

Data obtained were pooled and presented as means (±SEM). Data from 'control' mice were used as baseline values. The mean reaction times to the pain stimulus were recorded and subsequently analyzed using a 2-way ANOVA. Wilcoxon and Kruskal-Wallis tests were used to assess any differences. Statistical significance was obtained by using a double tailed CI of 95%, and a P value of less than 0.05. Pearson correlation coefficient was used to assess the activity of the extract compared to that of the synthetic capsaicin as shown in Figures 4.1-4.3.

1.3. RESULTS

1.3.1. Analgesic Activity

Capsicum frutescens-derived capsaicin extract (0.5–8.0 mg/kg i. p.) produced dose-dependent and significant (P<0.05–0.001) analgesic effects against thermally- and chemically-induced nociceptive pain (see Figures 4.1-4.3). Pretreatment of the mice with either CFE or CPF (0.5–5.0 mg/kg i. p.) caused significant (P<0.01–0.0002) delays in the mean reaction times of the animals to thermally-induced pain. (Figure 3.1). CFE and CPF (0.5-8.0 mg/kg i.p.) also caused dose-dependent and significant (P<0.05–0.001) reductions in the acetic acid-induced writhes of the mice (Figure 3.2).

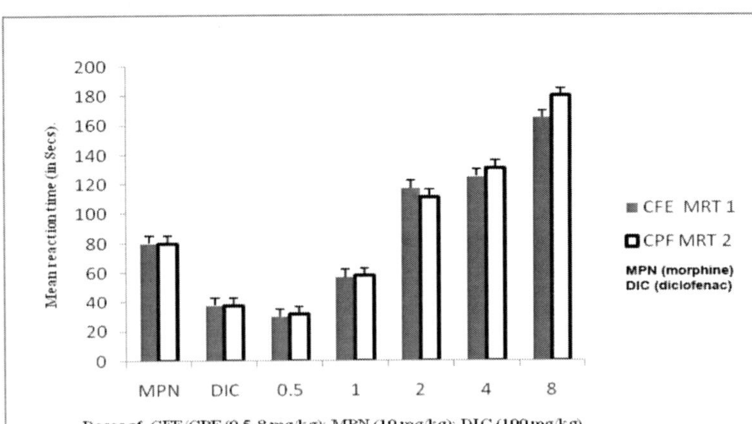

Figure 1. Mean reaction time (secs) to thermal pain following treatments of rats with CFE or CPF (0.5-8mg/kg) n=12 P< 0,01.

1.3.2. Activity Level in Mice

A general observation of the effects of capsaicin on the activities, sleep, pain tolerance and mechano-thermal sensation of mice, were noted. These effects were dose dependent, and wear off with time. The animals' quality of sleep was also better. There were initial excitation following injections of capsaicin, but this was followed by playful acts within each cohort, followed by calmness. These occur for both groups on reference and extracted capsaicin.

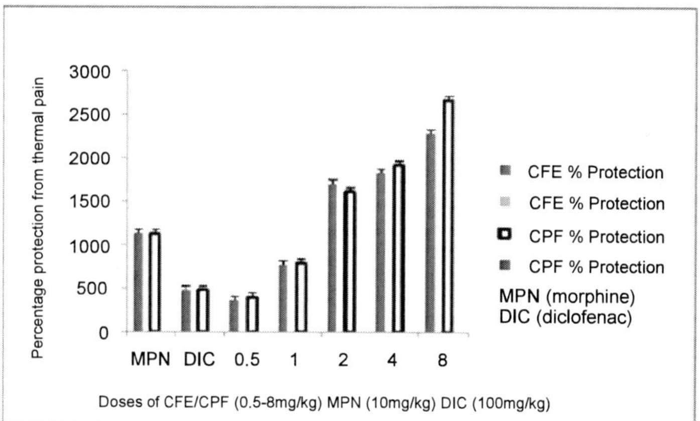

Figure 2. Percentage protection from thermal pain following treatment of rats with CFE or CPF (0.5-8 mg/kg) n=12 P<0,01.

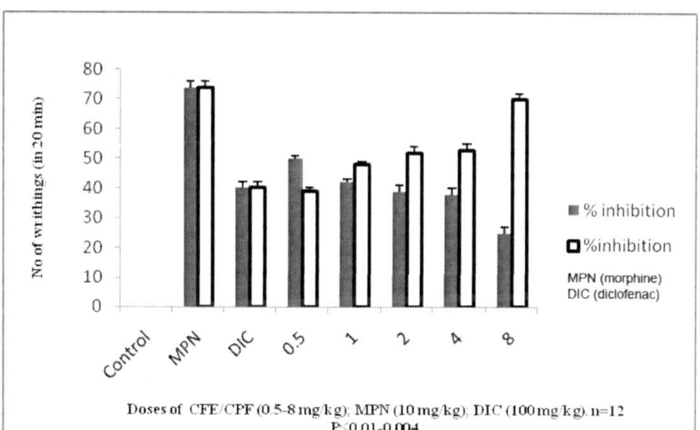

Figure 3. Doses of CFE/CPF and writhings (in 20 mins).

1.4. Discussion

Vanilloid 1 receptor has been described as the molecular integrator of several nociceptive stimuli (Ferreira et al., 2004). Most studies have alluded to the effectiveness of capsaicin-induced central analgesia. The present study aimed to explore the potential benefits of capsaicin in both central and peripheral analgesia.

The use of ‘acetic acid’ test method to investigate peripheral analgesic mechanism was validated at different times by Koster et al., (1959); Williamson et al. (1996); Zakaria et al., (2001); and Silva et al., (2003). Similarly, the evaluation of central analgesic mechanism using the ‘hot plate’ test method was validated by Eddy and Leimback, (1953); Lanhers et al., (1992); and Williamson et al., (1996). The reduction in the writhings of the abdominal wall and the prolongation of the mean reaction time to thermal pain can be regarded as indicative of positive peripheral and central analgesic mechanism, respectively.

Centrally-acting neuromediators can be classified into excitatory and inhibitory neuromediators. Glutamate and aspartate are examples of excitatory amino acids acting as neurotransmitters centrally. Substance P (SP), calcitonin-gene-related peptide (CGRP) and growth factors (e.g., brain derived neurotrophic factors) are other examples (Cervero, 1983; Chiang, et al., 1997, and 1999). Inhibitory neuromediators include endogenous opioids, such as enkephalin and β-endorphins. Others are gamma aminobutyric acid (GABA), glycine and α-adrenergic agonists. The effects of capsaicin at the GABAergic terminals, central MAO receptors, NMDA, acetylcholine and noradrenergic receptor sites were earlier elucidated (Davis, 1995; Davis, 2000; Woolf, 2000; 2001). The finding suggests that capsaicin activity at vanilloid receptors results in analgesia and cerebral depression at low doses, and that it produces paradoxical effects at other dissimilar receptors at high doses (Davis and Marinelli, 1993; Ralevic, 2001).

The aberration of inflammatory and neuropathic enhancement of pain perception as seen in allodynia (painful touch) and hyperalgesia are due to increased release of SP from the *substantia gelatinosa*. This phenomenon is called peripheral sensitization, and three principal neurotransmitters (NMDA, GABA, and Substance P) have been implicated in previous studies. Conversely, any agent acting on these receptors and neuromediators have the ability to modulate pain (Wallace, 1997).

Capsaicin showed a potent analgesic potential in the animals used. The explanation for the statistically often less significant effects at lower doses might have to do with the release of Substance P, due to de-vesiculation mechanism of receptor modification, receptor up-regulation and internalization (Yakish, 1999).

One of the findings of this study is that capsaicin has statistically significant ($P<0.002$) analgesic effects, especially for chronic therapy on centrally-mediated pain mechanisms. Peripherally-induced analgesia is statistically significant ($P< .001$) as compared to the 'control' at $P<0.5$. This finding probably suggests that capsaicin inhibits the inflammatory limb of pain mechanisms, or do so by receptor modification, and/or by the 'wind-up' mechanisms shown in previous studies (Winter, 1995).

The present study also shows that capsaicin has dose-dependent and statistically significant analgesic effects on mechano-thermal and chemically-induced pain. These results corroborate earlier studies that capsaicin is efficacious in neuropathic pain from diabetes, herpes, phantom and stump pain, chronic pain from osteoarthritis, and trigeminal neuralgia (Deal, 1991; Dray, 1990; 1992; Dini, 1993; Winter, 1995).

Furthermore, the stimulant effect of capsaicin results in convulsion and death at high doses (Nagy, 1980; 1983; Gade, 1985). Further work will be required to elucidate the mechanisms of this paradoxical effect.

CONCLUSION

Both capsaicin and *Capsicum frutescens* Linn [Solanaceae] fruit aqueous extract (CFE) have dose-dependent, statistically significant peripheral and central analgesic properties. More studies, however, need to be done towards evolving structural changes in their chemistry so as to develop capsaicin analogues adaptable for human uses in intractable pain from sports injuries, cancer management, and other forms of acute or chronic pain.

REFERENCES

Bevan, S. and Szolcsanyl J. (1990). Sensory neuron-specific actions of capsaicin: Mechanisms and applications. *Trends in Pharmacological Sciences;* 11(8):331-333.

Bevan, S. and Winter J. (1995). Nerve growth factor differentially regulates the chemosensitivities of cultured adult rat dorsal root ganglion neurons. *Journal of Neuroscience*; in press.

Cervero, F. and McRitchie, H.A. (1983). Effect of neonatal administration of capsaicin in several nociceptive systems of the rat. In: Bonica J.J. Liebeskind, J.C. and Albe-Fessard, D.G. (eds) *Advances in Pain Research and Therapy*. New York: Raven Press; 4:87-94.

Chiang, C.Y. Hu, J.W. and Sessle, B.J. (1997). NMDA receptor involvement in neuroplastic changes induced by neonatal treatment in trigeminal nociceptive neurons. *Journal of Neurophysiology*; 78(5): 2799-2803.

Chiang, C.Y. Kwan, C.L. Hu, J.W. and Sessle, B.J. (1999). Effects of GABA receptor antagonist on trigeminal caudalis nociceptive neurons in normal and neonatal capsaicin-treated rats. *American Physiological Society;* 82(5): 2154-2162.

Coderre, T.J. and Melzack, R. (1991). Central neural mediators of secondary hyperalgesia following heat injury in rats: Neuropeptides and excitatory amino acids. *Neuroscience Letters*; 131: 71-74

Davis, A. and Perkins, M.N. (1994). The involvement of bradykinin B1 and B2 receptor mechanisms in Substance P-induced mechanical hyperalgesia in the rat. *British Journal of Pharmacology*; 113: Suppl. 26P.

Davis, A. and Perkins, M.N. (1995). Capsaicin-induced mechanical hyperalgesia in the rat: involvement of NK1, bradykinin B1 and B2 receptors. *British Journal of Pharmacology*; 115: 684-688.

Davis, J.B. Gray, J. Gunthorpe, M.J. Hatcher, J.P. Davey, P.T. Overend, P. Harries, M.H. Latcham, J. Clapham, C. Atkinson, K. Hughes, S.A. Rance, K. Grau, E. Harper, A.J. Pugh, P.L. Rogers, D.C. Bingham, S. Randall, A. and Sheardown, S.A. (2000). Vanilloid receptor-1 is essential for inflammatory thermal hyperalgesia. *Nature;* 405 (6783): 183-187.

Dini, D. Bertelli, G. Gozza, A. and Forno, G.G. (1993). Treatment of postmastectomy pain syndrome with topical capsaicin. *Pain*; 54: 223-226.

Ferreira, J., da Silva, G.L., Calixto, J.B. (2004) Contribution of vanilloid receptors to the overt nociception induced by B2 kini receptor activation. *British Journal of Pharmacology*; 141:787-797.

Jaiarj, P. Saichompoo, S. Wongkrajang, Y. Vongswan, N. Peungvicha, P. and Jiratchariyakul, W. (1998). Cardiovascular actions of capsaicinoids extract from Thai Capsicum. *Thai Journal of Phytopharmacy;* 5(2): 1-13.

Jolayemi, A.T. (2002). Adverse drug reactions to pain medications in attendants of Frank Martin pain clinic, Addington Hospital. 1995-2001. *Regional Anaesthesia and Pain Medicine.*

Koster, R.M. Anderson J. and De Beer, E. J. (1959). Acetic acid for analgesic screening. *Federation Proceedings*; 18**:** 412–418.

Lanhers, M.C. Fleurentin, J. Dorfman, P. Mortier, F. Vinche, A. and Younos, C. (1992). Anti-inflammatory and analgesic effects of aqueous extract of harpagophytum procumbens. *Planta Medica*; 58 (2): 117-123.

Marinelli, S. Di Marzo, V. Berretta, N. Matias, I. Maccarrone, M. Bernardi, G. and Mercuri, N.B. (2003). Presynaptic facilitation of gluterminergic synapses to doperminergic neurons of the rat substantia nigra by endogenous stimulation of vanilloid receptors. *Journal of Neuroscience;* 23: 3136-3144.

Nagy, J.I. Vincent, S.R. Staines, W.A. Fibiger, H.C. Reisner, T. P. and Yamamura, H. I. (1980). Neurotoxic action of capsaicin on spinal Substance P neurons. *Brain Research*; 186 (2): 435-444.

Nagy, J.I. and van der Kooy, D. (1983). Effects of neonatal Capsaicin treatment on nociceptive thresholds in the rat. *Journal of Neuroscience*; 3: 1145-1150.

Nagy, J.I. (1985). Capsaicin action on the nervous system. In: Neurotransmitters in action (Bousfield D). *Elsevier Biomedical Pres-Amsterdam Trends in Neurosciences*; 180-187.

Ojewole, J. A., (2002). Anti-inflammatory properties of *Hypoxis hemerocallidea* corm ('African potato') extracts in rats. *Experimental Clinical Pharmacology*; 24 (10): 685-687.

Ralevic, V. Kendall, D.A. Jerman, J.C. Middlemiss, D.N. and Smart, D. (2001). Cannabinoid activation of recombinant and endogenous vanilloid receptors. *European Journal of Pharmacology*; 424 (3): 211-219.

Silvia, J. Abebe, W. Sousa, S.M. Duarte, V. G. Machado, M.I.L. and Matos, F.J.A. (2003). Analgesic and anti-inflammatory effects of essential oils of eucalyptus. *Journal of Ethnopharmacology*; 89 (2-3); 277-283.

Wallace, M.S. (1997). Advances in pain research. *Anaesthesiology Clinics of North America.*; 15(2): 229-234.

Watt, J.M. and Breyer-Brandwijk, M.G. (1962). *The medicinal and poisonous plants of Southern and East Africa*: 862-942.

Winter, J. Bevan, S. and Campbell, E.A. (1995). Capsaicin and pain mechanisms. *British Journal of Anaesthesia* 75: 157-168.

Woolf, C.J. and Max, M.B. (2001). Mechanism-based pain diagnosis: Issues for analgesic drug development. Anaesthesiology (in press)

Yakish, T.L. (1999). Spinal systems and pain processing: Development of novel analgesic drugs with mechanistically defined models. *Trends in Pharmacologicol Sciences;* 20: 329-337.

Zakaria, M. N. M. Islam, M. W. Radhakrishnan, R. Chen, H. B. Kamil, M. Al-Gifri, A. N. Chan, K. and Al-Attas, A. (2001). Antinociceptive and anti-inflammatory properties of caralluma arabica. *Journal of Ethnopharmacology;* 76 (2): 155-158.

In: Capsaicin
Editor: Bruce Gilliam
ISBN: 978-1-61470-433-1

Chapter 5

COMPARATIVE ANTI-INFLAMMATORY PROPERTIES OF CAPSAICIN AND ETHYL-ACETATE EXTRACT OF *CAPSICUM FRUTESCENS* LINN [SOLANACEAE] IN RATS

***A. T. Jolayemi*[*1] *and J. A. O. Ojewole*[2]**
[1]Department of Anaesthesia and ICU,
Goulburn Base Hospital and Senior Lecturer Australia National University (ANU) Goulburn Campus, Goulburn City NSW Australia
[2]Department of Pharmacology Faculty of Health Sciences, University of KwaZulu Natal, Durban, South Africa

ABSTRACT

Background: The analgesic effect of capsaicin (the active ingredient in *Capsicum frutescens* Linn. *[*Solanacea*e]*) had been reported in several studies. Current research is being directed at producing analgesics, anti-inflammatory agents with better side effect profile.

Objectives: To investigate if either the ethyl acetate extract of *Capsicum frutescens* Linn. *[*Solanacea*e]* (CFE) or capsaicin (Fluka Biotechnika-CPF) (in addition to the known analgesic properties) has any

* Corresponding author: Dr Adebayo Taiwo Ezekiel Jolayemi MBChB FMCA (Nig) FCA (SA) FANZCA PhD. adebayoj@hotmail.com, Department of Anaesthesia and ICU Goulburn Hospital NSW, Senior Lecturer, Australia, National University.

anti-inflammatory effect comparable to non-steroidal anti-inflammatory analgesics (NSAIDS).

Methods: The effects of ethyl acetate extract of *Capsicum frutescens* Linn. *[*Solanacea*e]* (CFE) and capsaicin (Fluka Biotechnika-CPF) was examined on rat hind paw. Inflammation was induced in the rat's hind paw by subplantar injections of fresh egg albumin (0.5 ml/kg). Diclofenac (100 mg/kg) was used as the reference anti-inflammatory agent for comparison, while distilled water was used as the placebo. The leucocytes count, corticosterone and C - reactive protein (CRP) levels were measured as biomarkers of inflammation. Data obtained were pooled and analysed using repeated ANOVA, in a general linear model with the CPSS software.

Results: Sub-plantar injections of fresh egg albumin (0.5 ml/kg) produced profound and time-related oedema in the rat hind paw of the 'control' rats. Diclofenac (DIC, 100 mg/kg, i.p.) and reference capsaicin (CPF, 2.5 mg/kg, i.p.) significantly inhibited paw swelling at ($p \leq 0.05$–0.001) (CI 95%) compared to distilled water-treated 'controls'. While the corticosterone levels were all very low in 7 rats treated with capsaicin, the leucocytes count was within normal range in 9 rats. However, in 16 specimens randomly assigned for CRP levels, there were very high CRP readings, up to a magnitude of 10 times the normal range.

Conclusion: Capsaicin in both forms (CFE and CPF) produced anti-inflammatory effects that were comparable to diclofenac in the experimental rat model at $p<0.05$. It may be concluded that capsaicin has both analgesic and anti-inflammatory properties.

Keywords: *Capsicum frutescens, 'chili',* ethylacetate extract, capsaicin, anti-inflammation

1.1. Introduction

Current anti-inflammatory agents have unacceptable systemic side effects. With the withdrawal of COX-2 agents such as rofecoxib (Vioxx) (Mandell, 2005; Giaquinta, 2004; Nelson et al., 2004); bextra, (Scheen, 2004); and the recommendation for a strong warning on labels of celecoxib and other remaining COX-2 agents (FDA, 2005); it is evident that natreceuticals and phytochemicals will be sought after as alternative anti-inflammatory and analgesic agents.

Prostaglandin inhibitors inhibit cyclo-oxygenase enzymes; leukotriene formation and pyrogen release (Capone et al., 2003). This makes their use to compromise gastric mucosal integrity and renal blood flow, resulting in peptic ulcer, upper gastro-intestinal tract (GIT) bleeding, as well as renal failure.

Current research reports have shown that capsaicin exhibits various potent biological activities following oral or topical administration. These include initial hyperalgesia, analgesia, dyspepsia, anti-inflammatory, anti-pyretic and anti-coagulatory activities. Capsaicin can modulate endocrine and paracrine activities, immune responses, as well as gastro-intestinal and cardiovascular functions.

Chronic inflammatory conditions were found to be associated with an up-regulation of Substance P receptors (De et al., 1990; Velazquez et al., 2002). Stimulation of transient receptor potential vanilloid 1 results in the activation of nociceptive and neurogenic inflammatory responses (Rigoni et al., 2003).

Capsaicin has proved very useful in intractable pain of inflammatory origin, such as from osteoarthritis, pruritus, psoriasis, rheumatism, and Complex Regional Pain Syndrome (Winter et al., 1995; Woolf, 1998; Yakish 1999; Yoshimura et al., 2000). Although *Capsicum frutescens* is traditionally used for the treatment of 'fever' (inflammation), the mechanism of its action is not well known.

One of the main objectives of this study was to examine if capsaicin of South African origin has any anti-inflammatory action. The use of carregenin-induced paw oedema to investigate anti-inflammatory properties of pharmacological substances has been reviewed by Winter et al., (1962) and cited by Winter et al. (1995). The increase in paw size following injection of carregenin, or egg albumin as a phlogistic agent, is regarded as indicative of inflammation (Ojewole, 2002; 2004; Saidu et al. 2000). The degree of inflammation is the difference between the initial paw size and the paw size at a chosen time interval. The percentage inflammation is thereafter calculated. The serum C - reactive protein (CRP) level was used to validate the extent of the inflammation produced by fresh egg albumin.

1.2. Materials and Methods

1.2.1. Fresh Egg Albumin-Induced Paw Oedema in Rats

Adult rats of both sexes weighing 250-300 g were randomly selected into 4 groups (A-D) of 10 rats per group. Baseline measurements of the circumferences of both the right and left hind limbs were obtained. Each rat in the control Group A received 10 ml/kg i.p. of normal saline, while each rat in

Group B received 100 mg/kg diclofenac (i.p.), respectively. Groups C and D rats received 2.5 mg/kg CFE and CPF, respectively. Each of the 40 rats was then subjected to inflammation by injecting fresh egg albumin (0.5 ml/kg) into the sub-plantar of the right paw according to standard protocol. Thereafter, the circumferences of the paws were repeatedly measured at time intervals of 30 min, 1, 2, 4, and 8 hours, respectively.

1.2.2. Biochemical Markers of Inflammation

Blood levels of leucocytes, C-reactive protein and cortisol were measured following prolong exposure to both forms of capsaicin and the phlogistic agent by intracardiac puncture under intravenous anaesthesia with midazolam ('hypnoval').

1.3. Results

1.3.1. Fresh Egg Albumn-Induced Paw Oedema

Sub-plantar injections of fresh egg albumin (0.5 ml/kg) produced profound and time-related oedema in the rat hind paw of the 'control' rats. Plantar swelling and/or oedema (which became evident approximately 20-30 minutes following fresh egg albumin administration) reached its peak approximately 90 minutes after the sub-plantar injection of fresh egg albumin. Diclofenac (DIC, 100 mg/kg) and reference capsaicin (CPF) signficantly inhibited paw swelling at ($P<0.05$–0.001) (CI 95%) compared to saline-treated 'controls'.

1.3.2. Capsaicin and Chemical Markers of Inflammation

Corticosterone levels were all very low (28-31 μmol/L) in 7 rats treated with capsaicin. Leucocytes count was within normal range (2.8-11.8 x 10^6/L) in 11 rats due to the innate immunity of the animals, and their intact defence mechanisms. In 16 specimens randomly assigned for CRP levels, there were very high readings, up to a magnitude of 10 times the normal (3-33 μmol/L).

Table 1.1. Comparative degree of anti-inflammatory effects of capsaicin in rats treated with phlogistic agent compared with parallel groups treated with diclofenac and saline at the same successive time intervals

GROUPS	Saline	Albumin *	Alb. + CFE**	Alb.+DIC***	Alb. + CPF***
%CO/Ct1	94 ± 1	96 ± 1	88 ± 1	86 ± 1	93 ± 1
%CO/Ct2	97 ± 1	96 ± 1	85 ± 1	83 ± 1	85 ± 1
%CO/Ct3	100 ± 1	103 ± 1	83 ± 1	80 ± 1	81 ± 1
%CO/Ct4	100 ± 2	103 ± 1	97 ± 2	95 ± 1	85 ± 1
%CO/Ct5	102 ± 2	103 ± 2	98 ± 2	96 ± 1	95 ± 1

*P<0.02 ** P<0.001 *** P<0.0001 using repeated ANOVA.

(Where C0= initial paw circumference; C1= paw circumference at 30 min; C2=paw circumference at 1 hr; C3= paw circumference at 2 hours; C4= paw circumference at 4 hours; C5= paw circumference at 8).

Table 1.2. Comparative degree of anti-inflammatory effects of capsaicin in rats treated with diclofenac and saline

GROUPS	Saline	Albumin*	Alb. + CFE**	Alb.+DIC***	Alb. + CPF***
%CO-Ct/C01	-4.1	-0.86	-12.43	-11	-16
%CO-Ct/C02	-1.98	-0.04	-0.97	-0.6	-21
%CO-Ct/C03	2.8	0	0.42	-24	-28
%CO-Ct/C04	3.6	1.43	0.23	-11	-8
%CO-Ct/C05	3.7	4.31	2.5	-5	-4

**P<0.001. *** P<0.0001. (Using repeated ANOVA at 95% CI).

Table 1.3. C-reactive protein (CRP) levels.

	CRP uml/L	Gp 1	% increase CRP	Gp2	% increase CRP	Gp 3	% increase CRP	Gp 4	% increase CRP
Control	2.5± .2		0		0		0		0
Rats		15±1	500±46	15±1	512±50		0		0

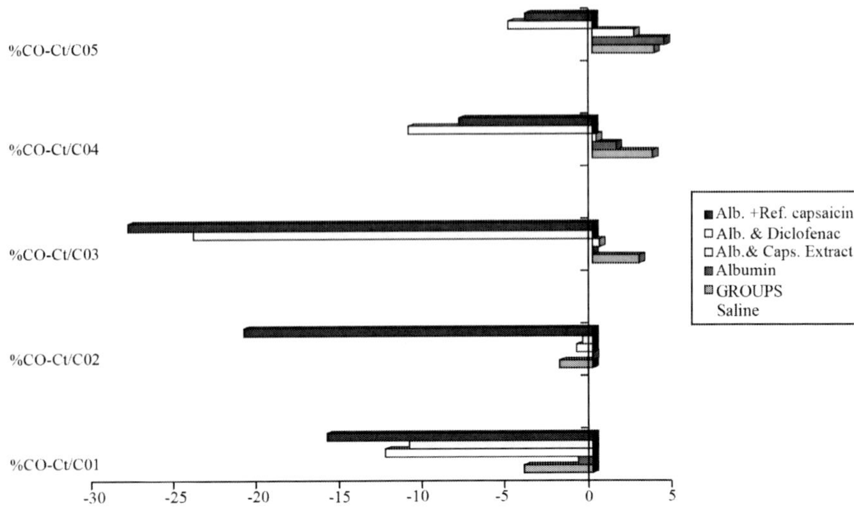

Figure 1.1. Degree of inhibition at 95% CI.

1.4. Discussion

The association of vanilloid receptors with inflammation and pain has been reviewed (Winter et al., 1995). It has been established that inflammation is a final pathway in chronic pain, connective tissue diseases, and various systemic disorders. With increase in the levels of Substance P in inflammatory and neurogenic joint diseases (arthritis), topical or intra-articular injections of capsaicin have shown a significant improvement, as well as reduction in the level of inflammatory mediators (Winter et al., 1995; Woolf, 1998; Yakish, 1999; Yoshimura et al., 2000).

Perkins and Campbell (1992) used 6 mg/kg of intra-articular capsaicin to reverse mechanical hyperalgesia for several hours. In rheumatoid arthritis, the findings are mixed. While Deal and co-workers (1991) showed significant reduction in the level of pain intensity in 31 patients with rheumatoid arthritis of the knee following treatment with Zostrix as 0.025% for 4 weeks, McCarthy and McCarty (1992; 1994) did not observe improvement in 7 patients with rheumatoid hands using 0.75% capsaicin. However, Menkes et al. (1993) reported that application of capsaicin (0.75%) for 6 weeks produced a reduction in inflammatory mediators, including Substance P, in the synovial fluid of patients with rheumatoid arthritis.

In osteoarthritis, there is evidence to show an increase in the level of Substance P in patients (Marshall, 1990; Mapp, 1994). Furthermore, randomized, controlled trials have shown significant improvements in pain relief following treatment with capsaicin cream. In addition, a double blind trial (Ellis, 1993) showed that capsaicin is useful in psoriasis. Post-operative use of capsaicin to prevent upper airway obstruction following Ear, Nose and Throat surgery was recently reviewed (Sicuteri 1989; Kowalski, 1999).

Studies in humans have shown that pain of inflammatory origin, such as neuropathic pain (Wallace, 1997); post-herpetic neuralgia (Berring, 1990; Kurkccuoglu, 1990; Lee, 1994; Leis et al. 2003); complex regional pain syndrome type I (Sicuteri, 1989); trigeminal neuralgia (Fusco, 1991); oral neuropathic pain (Epstein, 1994); cluster headache (Santiciolli, 1987; Maggi, 1988); fibromyalgia (Marks, 1993), in acute or chronic conditions, such as osteoarthritis (McCarty, 1994) and rheumatoid arthritis (Deal, 1991), have been managed successfully with capsaicin.

In addition, capsaicin has been shown to relief pruritus in patients with psoriasis (Ellis, 1993); brachioradial pruritus (Goodless and Eaglstein, 1993); aquagenic pruritus (Goodless and Eaglstein, 1993); notalgia parasthetica, nodular prurigo, and pruritus produced in patients on haemodialysis (Brand, 1987).

In human volunteers, capsaicin treatment inhibited itch after histamine and allergen challenge. Itching is mediated by a subset of capsaicin-sensitive nociceptive neurons. Blockade of C- fibre conduction suppresses itching (Lou, 1992).

Although several studies have alluded to the usefulness of capsaicin in the above clinical scenarios, these studies lack statistical or clinical significance. Neither was capsaicin compared with placebo or positive controls. In the present study, capsaicin was compared to placebo (negative 'control' saline), and positive controls (diclofenac and morphine) respectively.

Statistical significance was found in animals that received diclofenac, CPF and CFE, compared to animals in albumin and saline alone groups, respectively.

Diclofenac, CFE and synthetic capsaicin groups showed significant inhibition of inflammation. With regard to the biochemical inflammatory markers, the high CRP levels showed that the phlogistic agent induced inflammation in rats.

The only plausible explanation for the low cortisol levels in all the randomly-sampled animals is that CFE and CPF inhibited the unnecessary stress responses, while not inhibiting the beneficial pro-inflammatory activities to defend the body from infection. This hypothesis tends to suggest that capsaicin could play a useful role in stress reduction and inflammation.

Conclusion

Capsaicin in both forms (CFE and CPF) produced anti-inflammatory effects that were comparable to diclofenac in the experimental rat model at $p<0.05$. This study shows that capsaicin possesses dose-dependent anti-inflammatory effects.

References

Berring, P. Arendt-Nielsen, I. and Soderberg, U. (1990). Argon lasere induced cutaneous sensory and pain thresholds in post-herpeutic neuralgia. Quantitative modulation by topical capsaicin. *Acta Dermatology Venerology*; 70: 121-125.

Brand, L.L. Berman, E. Schwen, R. Loomans, M. Janusz, J. Bonne, R. Maddin, C. Gardner, J. LaHamn, T. Farmer, R. Jones, L. Chiaabrando, C. and Fanelli, R. (1987). NE-19950: A novel, orally active anti-inflammatory analgesic. *Drugs in Experimental and Clinical Research*; 13: 259-265.

Capone, M.L. Tacconelli, S. Sciulli, M.G. and Patrignani, P. (2003). Clinical pharmacology of selective COX-2 inhibitors. *International Journal of Immunopathology and Pharmacology*; (2 Suppl): 49-58.

Deal, C.L. Schnitzer, T.J. Lipstein, E. Seibold, J.R. Stevens, R.M. Levy, M.D. Albert, D. and Renold, F. (1991). Treatment of arthritis with topical capsaicin: a double-blind trial. *Clinical Therapeutics*; 13: 383-395.

De, A.K. and Ghosh, J.J. (1990). Inflammatory responses induced by Substance P in rat paw. *Indian Journal of Experimental Biology*; 28(10): 946-948.

Ellis, C.N. (1993). A double-blind evaluation of topical capsaicin in pruritic psoriasis. *Journal of the American Academy of Dermatology*; 29: 438-442.

Epstein, J.B. and Marcoe, J.H. (1994). Topical capsaicin for treatment of oral neuropathic pain and trigeminal neuralgia. *Oral Surgery Oral Medicine Oral Pathology*; 77: 135-140.

Fusco, B.M. Geppetti, P. Fanciullacci, M. and Sicuteri, F. (1991). Local application of capsaicin for the treatment of cluster headache and idiopathic trigeminal neuralgia. *Cephalalgia;* 11: 234-235.

Giaquinta, D. (2004). Lessons learned after the withdrawal of rofecoxib. *Managed Care Interface*; 17(11):25-26, 46.

Goodless, D.R. and Eaglstein, W.H. (1993). Brachioradial pruritus: Treatment with topical capsaicin. *Journal of the American Academy of Dermatology;* 29: 783-784.

Kowalski, M.L. Dietrich-Milobedzi, A. Majkowska-Wojciechowska, B. and Jarzebska, M. (1999). Nasal reactivity to capsaicin in patients with seasonal allergic rhinitis during and after the pollen season. *Allergy;* 54(8): 804-810.

Kurkccuoglu, N. and Alaybeyi, F. (1990). Topical Capsaicin for psoriasis. *British Journal of Dermatology*; 123: 549-550.

Lee, J.J. and Gauci, C.A. (1994). Postherpetic neuralgia: Current concepts and management. *British Journal of Hospital Medicine*; 52(11): 565-570.

Leis, S. Weber, M. Isselmann, A. Schmelz, M. and Birklein, F. (2003). Substance P-induced protein extravasation is bilaterally increased in complex regional pain syndrome. *Experimental Neurology*; 183(1): 197-204.

Lou, Y.P. Franco-Cereceda, A. and Lundberg, J.M. (1991). Omega-conotoxin inhibits CRGP release and bronchoconstriction evoked by a low concentration of capsaicin. *Acta Physiologica Scandinavica*; 141: 135-136.

Maggi, C.A. Patacchini, R. Santiciolli, P. Giuliani, S. Delbianco, F. Geppetti, P. and Meli, A. (1988). Protective action of ruthenium red toward capsaicin desensitization of sensory fibers. *Neuroscience Letters*; 88: 201-205.

Mandell, B.F. (2004). The vioxx withdrawal: latest in the COX 2 contoversies. *Cleveland Clinics Journal of Medicine*; 71(11): 849-852.

Mapp, P. and Kidd, B. (1994). The role of Substance P in rheumatic disease. *Seminars in Arthritis and Rheumatism;* 23: Suppl 3, 3-9.

Marks, D.R. Rapoport, A. Padla, D. Weeks, R. Rosum, R. Sheftell, F. and Arrowsmith, F A. (1993). Double-blind placebo-controlled trial of intranasal capsaicin for cluster headache. *Cephalalgia;* 13: 114-116.

Marshall, K.W. Chiu, B. and Inman, R.D. (1990). Substance P and arthritis: Analysis of plasma and synovial fluid levels. *Arthritis and Rheumatism;* 33: 87-90.

McCarthy, G.M. and McCarthy, D.J. (1992). Effect of topical capsaicin in the painful osteoarthritis of the hands. *Journal of Rheumatology*; 19: 604-607.

McCarty, D.J. Csuka, M. McCarthy, G. and Trotter, D. (1994). Treatment of pain due to fibromyalgia with topical capsaicin: A pilot study. *Seminars in Arthritis and Rheumatism;* 23: (Suppl 3) 41-47.

Menkes, C.J. Renoux, M. Loussadi, S. Mauborgne, A. Bruxelle, J. and Cesselin, J. (1993). Substance P (SP) levels in synovial tissue and synovial fluid from rheumatoid arthritis (RA) and osteoarthritis (OA) patients. *Journal of Rheumatology*; 20: 714-717.

Nelson, M. and Cicuttini, F. (2004) Lessons on prescribing and drug safety from withdrawal of rofecoxib. *Australian Family Physician*; 33(12): 965-966.

Ojewole, J.A.O. and Adewunmi, C.O. (2004). Anti-inflammatory and hypoglycaemic effects of tetrapleura tetraptera (Taub) [fabaceae] fruit aqueous extract in rats. *Journal of Ethnopharmacology*; 95(2-3): 177-182

Ojewole, J.A.O. (2007). Analgesic, anti-inflammatory and hypoglycaemic effects of Rhus chirindensis (Baker F.) [Anacardiaceae] stem-bark aqueous extract in mice and rats. *Journal of Ethnopharmacology*; 113(2): 338-345.

Perkins, M.N. and Campbell, E.A. (1992). Capsazepine reversal of the analgesic action of capsaicin in vivo. *British Journal of Pharmacology*; 107: 329: 184-194.

Rigoni, M. Trevisani, M. Gazzieri, D. Nadaletto, R, Tognetto, M. Cremon, C. Davis, S. Camp, B. Amadesi, S. Geppetti, P. and Harrison, S. (2003). Neurogenic responses mediated by Vanilloid receptor-1 (TRPV1) are blocked by the high affinity antagonist iodo-resiniferatoxin. *British Journal of Pharmacology*, 138:977-985.

Saidu, K. Onah, J. Orisadipe, A. Olusola, A. Wambebe, C. and Gamaniel K. (2000). Antiplasmodial, analgesic, and anti-inflammatory activities of the aqueous extract of the stem bark of erythrina senegalensis. *Journal of Ethnopharmacology*; 71 (1-2): 275-280.

Santiciolli, P. Patacchini, R. Maggi, C.A. and Meli, A. (1987). Exposure to calcium-free medium protects sensory fibers by capsaicin desensitization. *Neuroscience Letters*; 80: 167-172.

Scheen, A.J. (2004). Withdrawal of rofecoxib (Vioxx): What about cardiovascular safety of COX-2 selective non-steroidal anti-inflammatory drugs? *Review Med ical Liege;* 59(10): 565-9.

Sicuteri, F. Fusco, B.M. Marabini, S. Caimpagnolo, V. Maggi, C.A. Geppett, P. and Fanciullaci, M. (1989) Beneficial effect of capsaicin application to the nasal mucosa in cluster headache. *Clinical Journal of Pain;* 5:49-53.

Velaquez, R.A. McCarson, K.E. Cai, Y. Kovacs, K.J. Shi, Q. Evensjo, M. and Larson, A.A. (2002) Upregulation of neurokinin-1 receptor expression in rat spinal cord by an N-terminal metabolite of Substance P. *European Journal of NeuroSciences*.; 16 (2):229-241.

Wallace, M.S. (1997). Advances in pain research. *Anaesthesiology Clinics of North America* 15 (2):229-234.

Winter, J. Bevan, S. and Campbell, E.A. (1995) Capsaicin and pain mechanisms; *British Journal of Anaesthesia* 75: 157-168.

Yoshimura, M. Yonehara, N. Ito, T. Kawai, Y. and Tamura, T. (2000). Effects of topically applied capsaicin cream on neurogenic inflammation and thermal sensitivity in rats. *Journal of the Peripheral Nervous System*; 82(2): 116-121.

Yakish, T.L. (1999). Spinal systems and pain processing: Development of novel analgesic drugs with mechanistically defined models. *Trends in Pharmacologicol Science*; 20:329-337.

In: Capsaicin
Editor: Bruce Gilliam

ISBN: 978-1-61470-433-1

Chapter 6

CAPSAICIN TREATMENT IN STUDYING OF PERIPHERAL PAIN PROCESSING

V. V. Porseva, V. V. Shilkin and P. M. Masliukov
Dep. Normal Physiology, Yaroslavl State
Medical Academy, Yaroslavl, Russia

ABSTRACT

Capsaicin (trans-8-methyl-N-vanillyl-6-nonenamide) is a naturally occurring alkaloid derived from plants of the genus Capsicum. TRPV1 (transient receptor potential vanilloid-1), a capsaicin receptor, is expressed in primary sensory neurons of dorsal root ganglia and vagal nerves. Heat and protons as well as capsaicin activate TRPV1 to induce the influx of cations, particularly Ca^{2+} and Na^{+} ions. Characteristic effects of capsaicin are the induction of a burning sensation after acute administration and the desensitization of sensory neurons after large doses and prolonged administration. Also, activation of capsaicin-sensitive terminals enhances release of neuropeptides, for example substance P and calcitonin gene-related peptide. The use of capsaicin in animal models is important in the study of both processing and modulation of pain signals. Systemic treatment with the capsaicin is known to establish a long-lasting depletion of certain populations of afferent neurons and nerve fibers. Capsaicin induced different morphologic changes when given in different doses and at different ages. Administration of capsaicin modified age-related changes in the number of sensory neurons, containing TRPV1-receptors, calcitonin gene-related peptide, substance P, neurofilament 200.

INTRODUCTION

Pain sensation (nociception) is an alarm system aiming to signal the presence of potentially or actually harmful stimuli. As a submodality of somatic sensation, it has been defined as a "complex constellation of unpleasant sensory, emotional and cognitive experiences provoked by real or perceived tissue damage and manifested by certain autonomic, psychological, and behavioral reactions" (Dubin, Patapoutian, 2010). Pain is described as having different qualities and temporal features depending on the modality and locality of the stimulus. Normally, nociception and the perception of pain are evoked only at pressures and temperatures extreme enough to potentially injure tissues and by toxic molecules and inflammatory mediators. These high threshold physical and noxious chemical stimuli are detected by specialized peripheral sensory neurons (nociceptors) (Millan, 1999).

Unfortunately, under some pathological conditions, nociceptive signaling might start to misrepresent the sensory inputs and, at times, might be completely independent of sensory stimulation causing pain which is no longer beneficial for the organism. It is estimated that over one-third of the world's population suffers from persistent or recurrent pain.

Moreover, persistent pain, even at moderate levels, is often highly invalidating compromising working ability and/or everyday life activities. Finally, clinicians have few, if any, consistently effective means of relief chronic pain devoid of severe side effects (Gascon, Moqrich, 2010).

NOCICEPTORS

Nociceptive neurons act as the first element in the pain sensory machinery transducing noxious stimuli and transmitting nociceptive information to the central nervous system.

Numerous studies have revealed the existence of distinct classes of nociceptor activated by noxious stimuli. Adequate stimuli include temperature extremes, intense pressure, and chemicals signaling potential or actual tissue damage (Dubin, Patapoutian, 2010).

Nociceptor activity does not always lead to the perception of pain.

The sense of pain requires peripheral information to reach higher centers and normally depends on the frequency of action potentials in primary afferents, temporal summation of pre- and postsynaptic signals, and central influences (Willis, Coggeshall, 2004).

Sensory ganglia consist of different groups of neurons (Porseva et al., 2013a,b; Masliukov et al., 2014). There are three categories of afferent neurons: large neurons, small neurons that contain neuropeptides, and small neurons that do not contain neuropeptides (Snider and McMahon, 1998). Peptide-containing neurons are immunoreactive (IR) for substance P (SP) or calcitonin gene-related peptide (CGRP). Small, non-peptide-containing neurons binding the isolectin B4 (IB4), from Griffonia simplicifolia (IB4), may also be involved in various types of pain states, including that following injury. Neurofilament 200 (NF200) is a 200-kDa protein that is expressed in neurofilament-rich neurons and is often used as a marker for myelination (Perry et al., 1991). NF200 is usually located in large neurons (Molliver et al., 1995; Snider and McMahon, 1998). Large NF 200 positive neurons transmit proprioception, and small NF 200 positive neurons transmit pain. (Perry et al., 1991). Some neurons contain calcium binding proteins: calretinin and calbindin. They were shown to be particularly abundant in neuron associated to sensory pathways in both central and peripheral nervous system (Andressen et al., 1993; Ambrus et al., 1998; Masliukov et al., 2014).

Nociceptors are heterogeneous by a variety of molecular criteria (Julius et al., 2001; Braz et al., 2005), but the functional significance of this heterogeneity is unclear. Electrophysiological recordings of unmyelinated primary afferent (C) fibers from dorsal root ganglia (DRG) show that most (70%) nociceptors are polymodal: they can be activated by multiple types of painful stimuli, such as mechanical or thermal (Perl, 1996; Lawson et al., 2008; Cavanaugha et al., 2009).

Acritical issue in such experiments, however, regards the cellular specificity and timing of administration of the markers/reagents used for ablation. For example, if ablation of a population of DRG neurons using a given marker caused deficits in behavioral sensitivity to multiple pain modalities (Abrahamsen et al., 2008; Basbaum et al., 2009), it could reflect the expression of this marker either in a homogeneous population of polymodal nociceptors or in multiple modality-specific subpopulations. Pharmacological ablation of the central projections of neurons that express the heat-sensitive channel TRPV1 (Caterina et al., 1997) caused a complete loss of heat pain sensitivity, without affecting responses to noxious mechanical or cold stimuli.

CAPSAICIN

Capsaicin is widely used in clinical and experimental pain studies. Capsaicin is a member of the vanilloid family of compounds, trans-8-methyl-N-vanillyl-6-nonenamide with a molecular wight 305,42. Like other vanilloids, capsaicin has a benzene ring and a long hydrophobic carbon tail with a polar amide group (Hayman and Kam, 2008; O'Neill et al., 2012).

It is now well known that capsaicin exert their effects by activating the TRPV1 receptor (Hayman and Kam, 2008; Reyes-Escogido et al., 2011). TRP family of receptors consists of different types of receptors associated with non-specific ion channels destined for transferring ions, mainly Ca^2+ and Na+ (Liu et al., 2001; Lai et al., 2004; Pecze et al., 2013). TRPs are one of the most extensively studied families of ion channels present in sensory neurons. The six subfamilies include TRPV, TRPC (canonical), TRPM (melastatin), TRPML (mucolipins), TRPA (ankyrin), and TRPP (polycystin). TRPM8 and -A1 are thought to be involved in cold sensing, whereas seven others are activated by heat, over a distinct range of temperatures: TRPV1–4, TRPM2, TRPM4, and TRPM5 (Venkatachalam and Montell, 2007).

TRPV1 receptors locate on membranes of primary afferents participated in the transduction of nociceptive signals and formation of neurogenic inflammation (Szallasi ye al., 2007; O'Neill et al., 2012). The TRPV1 receptor is a nonselective ligand-gated cation channel. It is activated by chemical stimuli, including capsaicin and noxious heat (>43°C), as well as being stimulated by protons (pH <5.2), endogenous lipids, and certain inflammatory mediators (Szallasi and Blumberg, 1999; Woodbury et al., 2004). Stimuli are detected and transduced through opening of the ion channel, which results in entry of cations such as Na+ and Ca2+ to the neuron; although the channel is nonselective, it has been shown to have a high preference for Ca2+ (Caterina and Julius, 2001; Holzer, 2008).

Neurophysiological and neurochemical studies have showed that capsaicin interacts with TRPV1 receptors and induces excitation followed by desensitization and morphological destruction in the large group of afferent C and Aδ fibers.

Simultaneously, capsaicin stimulate short-term release of neuropeptides from terminals of primary afferents is followed by blocking of synthesis and transport of neuropeptides (O'Neill et al., 2012). Neurotoxic action of capsaicin is explained by redundant intracellular accumulation of Ca2+ and Na+, leading to hydration, proteases activation, accumulation of toxins, degeneration and cell death (Fuchs et al., 2005; Jin et al., 2005).

Activation of the TRPV1 may cause both sensitization but may also result in desensitization of nerve cells. Desensitization is a dose-dependent phenomenon, whereby repeated application of low dose or a single high dose of capsaicin led to immediate desensitization, whereas single lower doses cause activation (LaMotte et al., 1991; Kennedy et al., 2010).

Fast desensitization first involves capsaicin binding of TRPV1, which not only activates afferent fiber and pain pathways but also results in release of neuropeptides CGRP and SP from the C fiber terminals. Once released, terminals are depleted and the nerve may become desensitized (Maggi and Meli, 1988; Simone et al., 1998). Voltage-gated ion channels are inactivated, which also results in a rapid, short-lasting, desensitization as further action potentials cannot be generated (Simone et al., 1998; Tominaga and Tominaga, 2005). It is suggested that the desensitization of fibers is a protective mechanism to inhibit excessive Ca2+ influx, which leads to excitotoxic cell death. Activation of TRPV1 and entry of Ca2+ ions to the nerves activates a number of enzymes, which may lead to sensitization.

However, a second effect is the activation of other Ca2+-dependent enzymes such as protein phosphatase 2B/calcineurin, which is likely to lead to tachyphyaxis (Docherty et al., 1996; Mohapatra and Nau, 2005). Calcineurin dephosphorylates TRPV1 and decreases sensitivity to agonists such as capsaicin (Mohapatra and Nau, 2005).

Systemic administration of CAPS results in the death of approximately 50% of sensory neurons in rat neonates (Holzer, 1991) and in a loss of approximately 17% in adult rats (Jancsó et al., 1985). RTX, an ultrapotent CAPS analogue, almost completely eliminates TRPV1- expressing afferent neurons in adult rats, when applied systemically (Chen, Pan, 2006).

Capsaicin Treatment

Systemic application of capsaicin in neonatal rats results in the death of small primary afferent neurons and the degeneration of unmyelinated fibers in a dose-dependent manner (Nagy et al., 1983; Hiura et al., 1999, 2002; Hiura, 2000). Initially, it was believed that capsaicin had no effect on adult primary sensory neurons (Fitzgerald, 1983; Jancsó et al., 1987).

However, since then it has been shown that systemic capsaicin exposure in adult rats reduces the number of C- and Aδ-type fibers (Schicho et al., 1999) and 17% of the neurons within adult lumbar dorsal root ganglia (Jancsó et al., 1985).

Systemic capsaicin treatment might play a central role in desensitization to capsaicin (Buck and Burks, 1986; Holzer, 2008). Nevertheless, there has always been a lingering doubt as to what degree this depletion of neuropeptides reflects desensitization as opposed to possible neurotoxicity.

These acute responses severely limit the initial dose of capsaicin that can be given for desensitization. Capsaicin needs to be given repeatedly in increasing doses, is sufficient to render the animals fully insensitive to chemically evoked pain for 1 to 3 months (Szallasi, Blumberg, 1999).

There are some doubts about the ability of primary sensory neurons to be damaged by capsaicin to regenerate afferent fibers (Gallaher et al., 2010). It has been shown that capsaicin treatment results in long-term losses in nociceptive fields and C-fiber conduction (Lynn, 1985; Chung et al., 1990). Opposite, it has been reported that systemic capsaicin administration results in increased fiber outgrowth in cultures of spinal ganglia (Winter et al., 1992).

Besides, capsaicin induced neurogenesis, which leads to a complete recovery of nodose ganglion neurons 60 days after systemic capsaicin treatment destroyed nearly 50% of the neuronal population (Czaja et al., 2008). The restoration of spinal viscerosensory afferents may arise from three sources: 1) regeneration of damaged C and Aδ type axons originating from capsaicin-sensitive neurons; 2) sprouting of uninjured, capsaicin-insensitive neurons; and 3) injury-induced neuronal replacement (Gallaher et al., 2010).

It is known that, depending on the goals of the study and the species and age of animals, different capsaicin doses (from 50 to 150 mg/kg) and ways of applying it (single, fractional) are used.

The experience of using capsaicin for studying the morphological and histochemical characteristics of neurons indicates the availability of a dose of 100 mg/kg at single subcutaneous injection into newborn rat pups (Brzozowski et al., 1996; Rumyantseva, 2001).

Our own experience of using capsaicin shows that, to reveal qualitative and quantitative changes in neurons, the dose of the drug in the single administration needs to be increased to 150 mg/kg (Porseva, 2013).

It has been reported that in adult rats, with regard to the depletion of peripheral SP/CGRP containing nerve fibers, the subcutaneous administration of 125 mg/kg capsaicin is most effective (Gamse et al., 1981; Holzer, 1991).

In this case, the total dose of capsaicin is given over 3 days, with 25 mg/kg (0.5 ml) on the first day, and 50 mg/kg (1 ml) on the second and third days (Brouns et al., 2003).

EFFECTS OF CAPSAICIN TREATMENT ON TRPV1-IMMUNOPOSITIVE AFFERENT NEURONS

In rats, sensory ganglia contain both TRPV1-immunopositive and TRPV1-immunonegative neurons. A small proportion of neurons in the thoracic sensory ganglion contain TRPV1 receptors (Ma, 2002; Banerjee et al., 2007). As the rats developed, the proportion of TRPV1-positive neurons increased to day 20 and remained unaltered to the end of the observation period (Porseva et al., 2012a). The group of TRPV1-positive neurons consists of small cells. TRPV1-negative cells have different size: small, intermediate, and large (Ma, 2002; Porseva et al., 2012a).

During development, the percentage of TRPV1-positive neurons increases 1.4-fold with age, the greatest increase in the proportion of TRPV1-positive neurons is observed in 30-day-old rats (Table 1). In newborn and older rats, the mean cross-sectional area of TRPV1-positive neurons is significantly smaller than that of TRPV1-negative neurons.

After capsaicin treatment, rats of all age groups also show both TRPV1-positive and TRPV1-negative neurons. Administration of capsaicin decreases the proportion of TRPV1-positive neurons by 20% in 10-day-old rats (Table 1). In more mature animals, the proportion of TRPV1-immunopositive neurons in ganglia was also smaller, by 12–20%.

Groups of TRPV1-positive neurons contained 1–3 cells, as in controls, located predominantly in the center of ganglia. Comparative analysis of images of ganglion neuron populations showed that as in controls, the group of TRPV1-positive neurons was homogeneous and consisted of small neurons, while the populations of TRPV1-negative neurons in ganglia were heterogeneous (Porseva et al., 2012a).

Analysis of the mean cross-sectional area of ganglion neurons in the experimental group showed that TRPV1- positive neurons were smaller throughout the observation period (Table 2). Capsaicin treatment is followed by the death of small neurons. In the population of TRPV1-positive neurons, this decrease was by a factor of 1.2, while the decrease in TRPV1-negative neurons was by a factor of 1.3 (Piper, Docherty, 2000; Porseva et al., 2012a).

Developmental changes in the mean cross-sectional area of cells after capsaicin had different developmental dynamics in both groups of neurons. Thus, mean cell area in the population of TRPV1-positive neurons was stable throughout the observation period.

Table 1. Percentage and cross-sectional area of TRPV1+ and TRPV1- DRG neurons in control and capsaicin-treated rats during the development

Parameter	Age, days	Control group		Experimental group	
		TRPV1-positive nerons	TRPV1-negative neurons	TRPV1-positive nerons	TRPV1-negative neurons
Relative content, %	10	30,0±1,8	70,2±5,4	10,1±2,9**	89,7±1,5**
	20	42,3±2,7*	58,2±4,2*	26,2±0,8*,**	74,0±5,2*,**
	30	47,1±2,6*	53,5±3,0*	27,7±1,9*,**	73,3±5,1*,**
	60	44,0±1,8*	56,3±3,5*	31,3±0,7*,**	69,2±4,4*,**
	90	41,3±2,7	59,1±4,1*	29,2±1,4*,**	71,3±45*,**
Mean cross-sectional area, μm^2	10	211±13	289±33	179±18	382±23**
	20	277±47	510±44*	166±15**	408±33**
	30	282±83	620±39*	207±21**	642±52*
	60	454±78*	744±55*	266±65**	888±70*,**
	90	416±74*	435±74	218±59**	488±30

*statistically significant differences in comparison with 10-day-old rats (P<0.05).
**statistically significant differences in comparison with control rats (P<0.05).

This distinguished this parameter from its behavior in the control group, where there were increases on observation days 60 and 90. Among the population of TRPV1-negative neurons, the developmental dynamics of changes in the mean cross-sectional area of cells consisted of a progressive increase to 60 days and a decrease on observation day 90, which was also characteristic for the developmental dynamics in the control group.

Analysis of the cell composition of TRPV1-positive neurons showed that in rats of the control group at 10, 20, and 30 days of age, more than 50% of cells were very small in size (Table 2). At ages 60 and 90 days, more than 50% of cells had areas from 301 to 600 μm^2 and were small in size. In the experimental group, TRPV1-positive neurons were very small at all ages. In rats from 30 days of age, the proportion of very small neurons decreased, while the proportion of small cells increased and remained at that level to the end of the experiment.

After capsaicin treatment, TRPV1-positive cells became larger with time; nonetheless, more than 50% of these cells were very small to the end of the observation period (Porseva et al., 2012a).

Table 2. Percentage of TRPV1+ and TRPV1- DRG neurons having different size classes in control and capsaicin-treated rats during the development

Neurons	Age, days	Group of animals	Size classes, μm²				
			Up to 300	301-600	601-900	901-1200	More than
TRPV1-positive	10	Control	87,4±0,6	12,6±0,6	0	0	0
		Experimental	95,7±1,1	4,3±1,1	0	0	0
	20	Control	75±5	23±4	2,0±0,7	0	0
		Experimental	98,3±0,29	1,7±0,29	0	0	0
	30	Control	77,3±0,6	22,7±0,6	0	0	0
		Experimental	87,3±0,5	11,1±0,4	1,6±0,24	0	0
	60	Control	37,2±0,9	62,8±0,9	0	0	0
		Experimental	66,5±0,19	32,0±0,4	1,5±0,19	0	0
	90	Control	36,4±0,9	63,6±0,9	0	0	0
		Experimental	67,8±0,3	32,2±0,3	0	0	0
TRPV1-negative	10	Control	70,7±0,5	29,3±0,5	0	0	0
		Experimental	55,3±0,7	44,7±0,7	0	0	0
	20	Control	10,4±0,5	28,0±0,6	43,0±0,4	14,3±0,5	4,3±0,3
		Experimental	26,2±2,2	47,0±3	26,8±1,9	0	0
	30	Control	11,2±0,6	38,8±0,6	43,2±0,5	4,8±0,4	2,0±0,3
		Experimental	14,2±0,4	47,5±0,4	24,0±0,6	8,2±0,4	6,1±0,3
	60	Control	31,4±0,5	42,6±0,5	15,0±0,4	6,6±0,4	4,4±0,4
		Experimental	10±4	42,4±2,8	30±3	12±4	5,5±1,5
	90	Control	38,6±0,4	41,4±0,4	15,0±0,3	5,0±0,3	0
		Experimental	34,6±1,2	40,4±1,2	19,2±0,4	5,8±0,4	0

Action of Capsaicin on Sensory Neurons Which Contain Neurofilament 200

Neurofilaments are the main structural component of the cellular skeleton. They are expressed in neuronal cells and are necessary for the functioning of axonal transport (Capano et al., 2001; Shadiack et al., 2001). Three neuron-specific proteins differing in molecular weight - light (68 kDa), medium (160 kDa), and heavy (200 kDa) - enter into a composition of neurofilaments. NF200 is a marker of large neurons of SG, having myelinated axons and responding to mechanical stimuli (Perry et al., 1991; Lawson, Waddell, 2001).

Thus, NF200 is contained in 45% of the neurons of lower thoracic and upper lumbar ganglia of sheep, 1% of the neurons of lower lumbar ganglia, and 30% of the neurons of the upper thoracic and lumbar ganglia of rats (Porseva et al., 2013b; Russo et al., 2010).

During development of the rats, the relative content of NF200+ neurons in the sensory thoracic and lumbar DRG changes at both levels (Table 3). Thus, in DRG, the proportion of NF200+ neurons in the first 20 days of age decreases and increases between 20 and 30 days. From one month to one year ages, the proportion of NF200+ neurons does not change.

NF200 in the cytoplasm of neurons in the sensory ganglia at both segmental levels were distributed uniformly, and the fluorescence intensity of the reaction product did not change with age; fibers were also fluorescent, pointing to their myelination.

After capsaicin treatment, the age-related quantitative dynamics of NF200+ neurons in the thoracic DRG in the experimental group was characterized by a significant decrease in the proportion of these cells at 10 days of age as compared with rat pups aged three days; the relative content of immunopositive neurons showed virtually no change in subsequent age groups. In the lumbar DRG, the content of NF200+ neurons also decreased significantly by 10 days age and did not change to the end of the observation period (Porseva et al., 2013b). In both the thoracic and lumbar DRG, there were significant reductions in the proportions of NF200+ neurons from 30 days of age to the end of the observation period as compared with controls.

Table 3. Percentage (%) of NF200+ neurons in in control and capsaicin-treated rats during the development

Age, days	Thoracic DRG		Lumbar DRG	
	Control	Experiment	Control	Experiment
3	20,6 ± 1,4	17,8 ± 1,4	23,6 ± 0,8	19,4 ± 1,2
10	13,5 ± 1,0*	13,3 ± 0,8*	16,6 ± 1,0*	10,1 ± 0,5*,**
20	16,6 ± 1,2	13,9 ± 0,4*	14,8 ± 0,6*	9,7 ± 0,5*,**
30	30,4 ± 1,7*	10,1 ± 0,9*,**	28,2 ± 1,4*	11,6 ± 0,9*,**
60	27,6 ± 1,5*	9,3 ± 0,8*,**	30,5 ± 1,8*	10,0 ± 0,9*,**
90	29,0 ± 1,8*	8,7 ± 0,8*,**	29,1 ± 1,7*	8,4 ± 0,7*,**
180	22,4 ± 1,2	8,0 ± 0,3*,**	25,4 ± 1,7	9,3 ± 1,0*,**
360	19,5 ± 1,2	10,6 ± 0,8*,**	22,0 ± 1,2	9,5 ± 0,8*,**

*statistically significant differences in comparison with 3-day-old rats (P<0,05).

**statistically significant differences in comparison with control rats (P<0,05).

Table 4. Cross-sectional area (μm^2) of NF200+ neurons in in control and capsaicin-treated rats during the development

Age, days	Thoracic nerve ganglia		Lumbar nerve ganglia	
	Control	Experiment	Control	Experiment
3	426 ± 13	402 ± 11	415 ± 11	406 ± 12
10	712 ± 19*	686 ± 15*,**	509 ± 14*	493 ± 14
20	870 ± 19*	959 ± 19*,**	842 ± 19*	1126 ± 33*,**
30	1018 ± 31*	1368 ± 31*,**	1034 ± 21*	1134 ± 32*
60	1245 ± 32*	1022 ± 25*,**	1368 ± 44*	995 ± 24*,**
90	1149 ± 37*	987 ± 24*	1258 ± 39*	1046 ± 35*
180	1041 ± 24*	1046 ± 37*	1209 ± 34*	1008 ± 28*,**
360	1340 ± 34*	1191 ± 33*,**	1431 ± 48*	1010 ± 25*,**

*statistically significant differences in comparison with 3-day-old rats (P<0,05).
**statistically significant differences in comparison with control rats (P<0,05).

At 3, 10, and 20 days of age, the proportions of immunopositive neurons were also smaller than in controls, but these differences were not significant in the thoracic DRG, in contrast to the significant changes in the lumbar DRG. Thus, modeling of afferentation deficit led to decreases in the populations of sensory NF200+ neurons at both levels, though there was a segmental difference – there was a greater decrease in the number of immunopositive neurons in the lumbar DRG at 360 days of age (Porseva et al., 2013b).

In the control group, the mean cross-sectional areas of NF200+ neurons increased over the one-year observation period (Table 4). In 10-day-old rats there were no statistically significant differences between control and capsaicin-treated rats. In 20-day-old rats, the mean cross-sectional areas of immunopositive neurons in the experimental group were significantly greater than that in controls. In 60-day-old and 360-day-old rats, NF200+ neurons were smaller in the experimental group than in the control group (Porseva et al., 2013b).

ACTION OF CAPSAICIN ON SENSORY NEURONS CONTAINIG NEUROPEPTIDES

Approximately 60% of all sensory ganglionic neurons are presented by cells with small diameters that have myelinated fibers of the Aδ-type and unmyelinated fibers of the C-type (McCarthy, Lawson, 1990).

It was found in electrophysiological studies that approximately 90% of these cells in mice, rats, monkeys, and humans function as nociceptors. Small cells, in accordance with their neurochemical phenotypes, are divided into two classes, peptidergic and nonpeptidergic (Snider, McMahon, 1998). Approximately half of them express CGRP or substance P, as well as p75 neurotrophic receptor and nerve gowth factor-specific receptor TrkA. The central processes of these neurocytes reach the surface plate of the posterior horn of the spinal cord, where many afferent terminals end. TrkA is detected in approximately 80% of the neurons of rat lumbar ganglia containing substance P (Kashiba et al., 1996).

Calcitonin gene-related peptide (CGRP) is a 37-amino acid regulatory neuropeptide that is generated by alternative processing of the calcitonin gene (Rosenfeld et al., 1983; Ma, 2004). It is one of the most abundant peptides in nerve tissue and is widely expressed in both peripheral and central neurons (van Rossum et al., 1997; Schicho et al., 2005; Yasuchika et al., 2005). CGRP is presented both in the central and peripheral nervous system.

CGRP-containing neurons belong to the class of cells having C, Aδ and Aα/β fibers (McCarthy, Lawson, 1990). There are two forms of CGRP, α-CGRP, mainly contained in Aδ and C-fibers of the trigeminal nerve and spinal ganglia neurons; the β-isoform of this peptide is located mainly in the intestinal nerve plexuses (Edvinsson, 2008). Histochemical studies of cerebral perivascular nerve endings and trigeminal ganglion cells showed that more than half of the neurons contain CGRP (Edvinsson et al., 1987; Tajti et al., 1999). CGRP is the most common neuropeptide, found in a subpopulation of primary afferent neurons that account for 40–50% of the ganglion neurons, most of which are nociceptors (Lawson, Waddell, 2001).

The quantity of CGRP-positive cells highly varies depending on the target organ (Bennett et al., 1996) and the segment of spinal cord (Ohtori et al., 2003), even between the members of the same species (Tonra, Mendell, 1998). In neurons innervating skin, CGRP is determined in 53%, and in sensory neurons directing the fibers to the bone in 23% of the cells. CGRP is detected predominately in neurons of small and medium size (Ivanusic, 2009). In the vagal ganglia of mammals, a higher quantity of CGRP-immunoreactive neurons is contained in the cranial sensory ganglia as compared with the caudal one. In cats the ratio of CGRP-immunoreactive neurons in these ganglia is 72 and 42%; in rats it is 30 and 10% (Franco-Cereceda et al., 1987; Torrealba, 1992; Ichikawa et al., 1993). In vagal sensory neurons approximately half of CGRP-immunoreactive neurons contains substance P; in turn, almost all of the substance P-immunoreactive neurons are CGRP-immunoreactive as well (Helke, Niederer, 1990; Ichikawa et al., 1991).

Substance P belongs to a family of neuropeptides known as tachykinins that share the common C-terminal sequence: Phe-X-Gly-Leu-Met-NH2. The three most common tachykinins are substance P, neurokinin A (NKA), and neurokinin B (NKB); their biologic actions are mediated through specific cell-surface receptors designated NK1, NK2, and NK3, with substance P the preferred agonist for NK1 receptors, NKA for NK2 receptors, and NKB for NK3 receptors (Saffroy et al., 1987; Shigemoto et al., 1990; Khawaja, Rogers, 1996). The demonstration of substance P immunoreactivity in the cell bodies of dorsal root ganglia, in sensory nerve fibers, and in the dorsal horn of the spinal cord led to early speculation that substance P is involved in pain perception (Zubrzycka, Janecka, 2000; Suzuki et al., 2002; Wiley et al., 2007).

During development, fractions of CGRP-immunoreactive neurons in the sensory vagal nodose ganglion (NG) and DRG changed differently. Until the age of 20 days, in the NG, the number of CGRP-immunoreactive neurons decreases, whereas, in the DRG, it increases. Then, approximately 10% of those cells in the NG and 27% of those cells in the DRG were observed at the next ages and remained stable in older animals (Porseva et al., 2012 b; Poyner et al., 2002; Aoki et al., 2005). A common feature of both ganglia is a stable quantitative composition of population CGRP-immunopositive neurons from 20 days to 6 months. Moreover, not only neuroplasm exhibited immunopositivity but also neuronal fibers, which is in accordance with data of other studies (Xu et al., 2005; Price et al., 2007). Population of CGRP-immunopositive neurons was heterogenous in the size of cells and, in adult rats, consisted of cells of all size classes from very small to very large; however, more than 50% of cells were small in size in both ganglia studied (Poyner et al., 2002; Xu et al., 2005; Price et al., 2007).

In addition, neurons in the NG had larger area as compared to DRG neurons, which is in accordance with the data on neuronal composition of the ganglia (Aoki et al., 2005; Bucelli et al., 2008).

Several studies have shown that capsaicin induced either massive loss of neurons independently on their sizes in the sensory ganglia (Torsney et al., 2000) or death of only small neurons, which are recognized as vulnerable to capsaicin (Piper et al., 2000; Ma et al., 2002).

Our study demonstrate that after capsaicin-induced chemical deafferentation, CGRP-immunopositive neurons were not observed in the NG, whereas, in the thoracic DRG, the population of these neurons decreased by 20 or 70% at the age of 3 days or other ages, respectively (Table 5).

Table 5. Percentage and cross-sectional area of CGRP+ DRG neurons in in control and capsaicin-treated rats during the development

Parameter	Age, days	NG		DRG	
		control	experiment	control	experiment
	3	20,0 ± 1,42	15,4 ± 1,48**	12,8 ± 0,62	10,5 ± 1,59**
	10	14,6 ± 1,34*	–	29,2 ± 1,48*	9,6 ± 0,84**
Percentage (%)	20	7,3 ± 1,29*	–	27,3 ± 1,75*	7,5 ± 0,17*, **
	30	6,4 ± 1,18*	–	27,9 ± 1,63*	8,5 ± 0,31**
	60	7,0 ± 1,22*	–	27,6 ± 1,36*	7,2 ± 0,24*, **
	90	6,6 ± 1,20*	–	26,3 ± 1,45*	7,6 ± 0,22*, **
	180	7,1 ± 1,17*	–	27,2 ± 1,64*	7,4 ± 0,16*, **
	3	349,2 ± 15,02	320,9 ± 14,67	321,6 ± 12,73	312,7 ± 13,85
Cross-sectional	10	419,5 ± 14,62*	–	303,8 ± 9,39	287,8 ± 12,20*
area (μm^2)	20	501,1 ± 43,78*	–	356,4 ± 12,33*	264,7 ± 12,09*
	30	657,4 ± 34,15*	–	373,8 ± 18,74*	290,3 ± 12,35*, **
	60	629,1 ± 15,73*	–	401,2 ± 14,12*	322,5 ± 9,17**
	90	623,3 ± 33,97*	–	462,7 ± 23,09*	348,6 ± 22,42**
	180	613,3 ± 16,97*	–	534,3 ± 25,98*	591,5 ± 23,37*, **

*statistically significant differences in comparison with 3-day-old rats (P<0,05).
**statistically significant differences in comparison with control rats (P<0,05).

Surviving positive neurons exhibited mainly granular staining in the neuroplasm. After capsaicin-induced chemical deafferentation, the population of CGRP-immunoreactive neurons was more homogenous and up to the age of 2 month it consisted of very small cells and at 3 and 6 months, it consisted of small cells with the lower mean area of neurons as compared to control.

After capsaicin-induced chemical deafferentation, CGRP-immunopositive neurons were not observed in the NG, whereas, in the thoracic DRG, the population of these neurons decreased by 20 or 70% at the age of 3 days or other ages, respectively. Surviving positive neurons exhibited mainly granular staining in the neuroplasm.

After capsaicin-induced chemical deafferentation, the population of CGRP-immunoreactive neurons was more homogenous and up to the age of 2 month it consisted of very small cells. In 3- and 6-month-old rats, it consisted of small cells with the lower mean area of neurons as compared to control.

Table 6. Percentage of CGRP+ DRG neurons having different size classes in control and capsaicin-treated rats during the development

Age (days)	Group	Up to 300	301–600	601–900	901–1200	More than 1200
3	Control	44,8 ± 1,36	55,2 ± 0,71			
	Experiment	53,7 ± 0,93	46,3 ± 0,86			
10	Control	58,0 ± 0,86	40,6 ± 1,03	1,4 ± 0,37		
	Experiment	55,5 ± 0,71	44,5 ± 0,71	–		
20	Control	44,0 ± 1,08	54,0 ± 0,86	2,0 ± 0,51		
	Experiment	57,7 ± 0,66	42,3 ± 0,75	–		
30	Control	40,0 ± 1,02	52,0 ± 0,93	8,0 ± 1,02		
	Experiment	59,2 ± 0,66	40,8 ± 0,81	–		
60	Control	30,0 ± 1,03	55,4 ± 0,66	9,3 ± 1,02	5,3 ± 0,66	
	Experiment	52,1 ± 1,17	43,9 ± 0,66	4,0 ± 0,66	–	
90	Control	16,5 ± 0,66	57,8 ± 1,23	8,2 ± 1,02	15,3 ± 0,66	2,2 ± 0,51
	Experiment	41,4 ± 1,17	51,0 ± 1,48	7,6 ± 0,63	–	–
180	Control	14,7 ± 0,66	61,8 ± 1,17	5,9 ± 0,63	14,7 ± 0,66	2,9 ± 0,63
	Experiment	–	50,0 ± 1,11	37,5 ± 0,95	12,5 ± 0,58	–

*statistically significant differences in comparison with 3-day-old rats (P<0,05).
**statistically significant differences in comparison with control rats (P<0,05).

In contrast to the control, during the first month of postnatal life, neurons of medium size were not found; during the second and third months, large neurons were not observed, and, at six months very small or very large cells were not found (Table 6) (Porseva et al., 2012 b).

Conclusion

After single capsaicin treatment, there is a cell death of sensory neurons with different sizes. Presumably, the maximal damage is observed in small cells, however, some large neurons can be also destroyed. Initial changes of neurons with separate neurochemical features are observed even in 10-day-old rats with maximal changes in 20-day-old animals. After that, existing population of sensory neurons remains constant in the first year of life.

In capsaicin treated rats, the cross-sectional area of sensory neurons was lower in comparison with control groups. In this case we can suggest that not only small, but some large neurons are capsaicin-sensitive. From the other hand, action of capsaicin induces neurogenesis.

As a result, morphological parameters of sensory neurons change after capsaicin treatment due to different types of neuronal damage and appearance of new neurons.

References

Abrahamsen, B., Zhao, J., Asante, C. O., Cendan, C. M., Marsh, S., Martinez-Barbera, J. P., Nassar, M. A., Dickenson, A. H., Wood, J. N. (2008). The cell and molecular basis of mechanical, cold, and inflammatory pain. *Science,* 321, 702-705.

Ambrus, A., Kraftsik, R., Barakat-Walter, I. (1998). Ontogeny of calretinin expression in rat dorsal root ganglia. *Brain Res. Dev. Brain Res.,* 106, 101-108.

Andressen, C., Blumcke, I., Celio, M. R. (1993). Calcium-binding proteins: selective markers of nerve cells. *Cell Tis. Res.,* 271, 181-208.

Aoki, Y., Ohtori, S., Takahashi, K., Ino, H., Douya, H., Ozawa, T., Saito, T., Moriya, H. (2005). Expression and co-expression of VR1, CGRP, and IB4-binding glycoprotein in dorsal root ganglion neurons in rats: differences between the disc afferents and the cutaneous afferents. *Spine,* 30, 1496-500.

Banerjee, B., Medda, B. K., Lazarova, Z., Bansal, N., Shaker, R., Sengupta, J. N. (2007). Effect of reflux-induced inflammation on transient receptor potential vanilloid one (TRPV1) expression in primary sensory neurons innervating the oesophagus of rats. *Neurogastroenterol. Motil.,* 19, 681-691.

Basbaum, A. I., Bautista, D. M., Scherrer, J., Julius, D. (2009). Cellular and Molecular Mechanisms of Pain. *Cell,* 139, 267-284.

Bennett, D. L., Dmietrieva, N., Priestley, J. V., Clary, D., McMahon, S. B. (1996). TrkA, CGRP and IB4 expression in retrogradely labeled cutaneous and visceral primary sensory neurones in the rat. *Neurosci. Lett.,* 206, 33-36.

Braz, J. M., Nassar, M. A., Wood, J. N., Basbaum, A. I. (2005). Parallel “pain” pathways arise from subpopulations of primary afferent nociceptor. *Neuron,* 47, 787-793.

Brouns, I., Genechten, J. V., Hayashi, H., Gajda, M., Gomi, T., Burnstock, G., Timmermans, J. P., Adriaensen, D. (2003). Dual sensory innervation of pulmonary neuroepithelial bodies. *Am. J. Respir. Cell Mol. Biol.,* 28, 275-285.

Brzozowski, T., Konturek, S. J., Sliwowski, Z., Pytko-Polończyk, J., Szlachcic, A., Drozdowicz, D. (1996). Role of capsaicin-sensitive sensory nerves in gastroprotection against acid-independent and acid-dependent ulcerogens, *Digestion,* 57, 424-432.

Bucelli, R. C., Gonsiorek, E. A., Kim, W. Y., Bruun, D., Rabin, R. A., Higgins, D., Lein, P. J. (2008). Statins decrease expression of the proinflammatory neuropeptides calcitonin gene-related peptide and substance P in sensory neurons. *Pharmacol. Exp. Ther.,* 324, 1172-1180.

Buck, S. H., Burks, T. F. (1986). The neuropharmacology of capsaicin: review of some recent observations. *Pharmracol. Rev.,* 38, 179-226.

Capano, C., Pernas-Alonso, R., Poryio, U. (2001). Neurofilament homeostasis and motoneurone degeneration. *BioEssays,* 23, 24-33.

Caterina, M. J., Schumacher, M. A., Tominaga, M., Rosen, T. A., Levine, J. D., Julius, D. (1997). The capsaicin receptor: a heat-activated ion channel in the pain pathway. *Nature,* 389, 816-824.

Caterina, M. J., Julius, D. (2001). The vanilloid receptor: a molecular gateway to the pain pathway. *Annu. Rev. Neurosci.,* 24, 487-517.

Cavanaugh, D. J., Lee, H., Lo, L., Shields, S. D., Zylka, M. J., Basbaum, A. I., Anderson, D. J. (2009). Distinct subsets of unmyelinated primary sensory fibers mediate behavioral responses to noxious thermal and mechanical stimuli. *Proc. Natl. Acad. Sci. US*, 106 9075-9080.

Chen, S. R., Pan, H. L. (2006). Loss of TRPV1-expressing sensory neurons reduces spinal mu opioid receptors but paradoxically potentiates opioid analgesia. *Neurophysiol.,* 95, 3086-3096.

Chung, K., Klein, C. M., Coggeshall, R. E. (1990). The receptive part of the primary afferent axon is most vulnerable to systemic capsaicin in adult rats. *Brain Res.,* 511, 222-226.

Czaja, K., Burns, G. A., Ritter, R. C. (2008). Capsaicin-induced neuronal death and proliferation of the primary sensory neurons located in the nodose ganglia of adult rats. *Neurosci.,* 154, 621-630.

Docherty, R. J., Yeats, J. C., Bevan, S., and Boddeke, H. W. (1996). Inhibition of calcineurin inhibits the desensitization of capsaicin-evoked currents in cultured dorsal root ganglion neurones from adult rats. *Pflugers Arch.,* 431, 828-837.

Dubin, A. E., Patapoutian, A. (2010). Nociceptors: the sensors of the pain pathway. *Clin. Invest.,* 120, 3760-3772.

Edvinsson, L. CGRP blockers in migraine therapy: where do they act? *Br. J. Pharmacol.* 155 (7): 967-969. 2008.

Edvinsson, L., Ekman, R., Jansen, I., et al. (1987). Calcitonin generelated peptide and cerebral blood vessels: distribution and vasomotor effects. *Cereb. Blood Flow Metab.*, 7, 720-728.

Fitzgerald, M. (1983). Capsaicin and sensory neurons - a review. *Pain,* 15, 109-130.

Franco-Cereceda, A., Henke, H., Lundberg, J. M., Petermann, J. B., Hökfelt, T., Fischer, J. A. (1987). Calcitonin gene-related peptide (CGRP) in capsaicin-sensitive substance P-immunoreactive sensory neurons in animals and man: distribution and release by capsaicin. *Peptides,* 8, 399-410.

Fuchs, A., Lirk, P., Stucky, C., Abram, S., Hogan, Q. (2005). Painful nerve injuri decreased resting cytosolic calcium concentrations in sensory neurons of rats. *Anesthesiology,* 102, 1217-1225.

Gallaher, Z. R., Larios, R. M., Ryu, V., Sprunger, L. K., Czaja, K. (2010). Recovery of viscerosensory innervation from the dorsal root ganglia of the adult rat following capsaicin-induced injury. *Comp. Neurol.,* 518, 3529-3540.

Gamse, R., Leeman, S. E., Holzer, P., and Lembeck, F. (1981). Differential effects of capsaicin pretreatment on the content of somatostatin, substance P, and neurotensin in the nervous system of the rat. *Naunyn Schmiedebergs Arch. Pharmacol.,* 317, 140-148.

Gascon, E., Moqrich, A. (2010). Heterogeneity in primary nociceptive neurons: from molecules to pathology. *Arch. Pharm. Res.*, 33(10), 1489-1507.

Hayman, M. and Kam, P. (2008). Capsaicin: a review of its pharmacology and clinical applications. *Curr. Anaesth. Crit. Care,* 19, 338-343.

Helke, C. J., Niederer, A. J. (1990). Studies on the coexistence of substancc P with other putative transmitters in the nodose and petrosal ganglia. *Synapse,* 5, 144-151.

Hiura, A. (2000). Neuroanatomical effects of capsaicin on the primary afferent neurons. *Arch. Histol. Cytol.,* 63, 199-215.

Hiura, A., Nakagawa, H., Koshigae, Y., Yoshizako, A., Kubo, Y., Ishizuka, H. (1999). Age-related changes in the response to thermal noxious heat and reduction of C-fibers by neonatal treatment with capsaicin. *Somatosens. Mot. Res.,* 16, 115-121.

Hiura, A., Nakae, Y., Nakagawa, H. (2002). Cell death of primary afferent nerve cells in neonatal mice treated with capsaicin. *Anat. Sci. Int.,* 77, 47-50.

Holzer, P. (1991). Capsaicin: cellular targets, mechanisms of action, and selectivity for thin sensory neurons. *Pharmacol. Rev.,* 43, 143-201.

Holzer, P. (2008). The pharmacological challenge to tame the transient receptor potential vanilloid-1 (TRPV1) nocisensor. *British J. Pharmacol.,* 155, 1145-1162.

Ichikawa, H., Rabchevsky, A., Helke, C. J. (1993). Presence and coexistence of putative neurotransmitters in carotid sinus baro- and chemoreceptor afferent neurons. *Brain Res.,* 611, 67-74.

Ichikawa, H., Jacobowitz, D. M., Winsky, L., Helke, C. J. (1991). Calretinin-immunoreactivity in vagal and glossopharyngeal sensory neurons of the rat: distribution and coexistence with putative transmitter agents. *Brain Res.,* 557, 316-321.

Ivanusic, J. J. (2009). Size, neurochemistry, and segmental distribution of sensory neurons innervating the rat tibia. *Com. Neurol.,* 517, 276-283.

Kashiba, H., Ueda, Y., Senba, E. (1996). Coexpression of preprotachykinin-A, alpha-calcitonin gene-related peptide, somatostatin, and neurotrophin receptor family messenger RNAs in rat dorsal root ganglion neurons. *Neurosci.,* 70, 179-189.

Kennedy, W. R., Vanhove, G. F., Lu, S. P., Tobias, J., Bley, K. R., Walk, D., Wendelschafer-Crabb, G., Simone, D. A., and Selim, M. M. (2010). A randomized, controlled, open-label study of the long-term effects of NGX-4010, a high-concentration capsaicin patch, on epidermal nerve fiber density and sensory function in healthy volunteers. *Pain,* 11, 579-587.

Jancsó, G., Kiraly, E., Joo, F., Such, G., Nagy, A. (1985). Selective degeneration by capsaicin of a subpopulation of primary sensory neurons in the adult rat. *Neurosci. Lett.,* 59, 209-214.

Jancsó, G., Kiraly, E., Such, G., Joo, F., Nagy, A. (1987). Neurotoxic effect of capsaicin in mammals. *Acta Physiol. Hung.,* 69, 295-313.

Jin, H. W., Ichikawa, H., Fujita, M., Yamaa, T., Mukae, K., Nomura, K., Sugimoto T. (2005). Involvement of caspase cascade in capsaicin-induced apoptosis of dorsal root ganglion neurons. *Brain Res.,* 1056, 139-146.

Julius, D., Basbaum, A. I. (2001). Molecular mechanisms of nociception. *Nature,* 413, 203-210.

Khawaja, A. M., Rogers, D. F. (1996). Tachykinins: receptor to effector. *Int. J. Biochem. Cell Biol.,* 28, 721-38.

Lai, J., Porreca, F., Hunter, J. C., Gold, M. S. (2004). Voltage-gated sodium channels and hyperalgesia. *Ann. Rev. Pharmacol. Toxicol.,* 44, 371-397.

LaMotte, R. H., Shain, C. N., Simone, D. A., and Tsai, E. F. (1991). Neurogenic hyperalgesia: psychophysical studies of underlying mechanisms. *Neurophysiol.,* 66, 190-211.

Lawson, S. N. and Waddell, P. J. (2001). Soma neurofilament immunoreactivity is related to cell size and fiber conduction velocity in rat primary sensory neurons. *Physiol.,* 435, 41-63.

Lawson, J. J., McIlwrath, S. L., Woodbury, C. J., Davis, B. M., Koerber, H. R. (2008). TRPV1 unlike TRPV2 is restricted to a subset of mechanically insensitive cutaneous nociceptors responding to heat. *Pain*, 9, 298-308.

Liu, L., Oortgiesen, M., Li, L., Simon, S. A. (2001). Capsaicin inhibits activation of voltage-gated sodium currents in capsaicin-sensitive trigeminal ganglion neurons. *Neurophysiol.,* 85, 745-758.

Lynn, B. (1987). The immediate and long-term effects of applying capsaicin to cutaneous nerves. *Acta Physiol. Hung.,* 69, 287-294.

Ma, H. (2004). Calcitonin Gene-Related Peptide (CGRP). *Nature and Science,* 2, 41-47.

Ma, Q. P. (2002). Expression of capsaicin receptor (VR1) by myelinated primary afferent neurons in rats. *Neurosci. Lett.,* 319, 87-90.

Maggi, C. A. and Meli, A. (1988). The sensory-efferent function of capsaicin-sensitive sensory neurons. *Gen. Pharmacol.,* 19, 1-43.

Masliukov, P. M., Emanuilov, A. I., Madalieva, L. V., Moiseev, K. Y., Bulibin, A. V., Korzina, M. B., Porseva, V. V., Korobkin, A. A., Smirnova, V. P. (2014). Development of nNOS-positive neurons in the rat sensory and sympathetic ganglia. *Neurosci.,* 256, 271-281.

McCarthy, P. W., Lawson, S. N. (1990). Cell type and conduction velocity of rat primary sensory neurons with calcitonin gene-related peptide-like immunoreactivity. *Neurosci.,* 34, 623-632.

Millan, M. J. (1999). The induction of pain: an integrative review. *Progress Neurobiol.,* 57, 1-164.

Mohapatra, D. P. and Nau, C. (2005). Regulation of Ca2-dependent desensitization in the vanilloid receptor TRPV1 by calcineurin and cAMP-dependent protein kinase. *Biol. Chem.,* 280, 13424-13432.

Molliver, D. C., Snider, W. D. (1997). Nerve growth factor receptor TrkA is down-regulated during postnatal development by a subset of dorsal root ganglion neurons. *Comp. Neurol.,* 381, 428-438.

Nagy, J. I., Iversen, L. L., Goedert, M., Chapman, D., Hunt, S. P. (1983). Dose-dependent effects of capsaicin on primary sensory neurons in the neonatal rat. *Neurosci.,* 3, 399-406.

Ohtori, S., Takahashi, K., Chiba, T., Yamagata, M., Sameda, H., Moriya, H. (2003). Calcitonin gene-related peptide immunoreactive neurons with dichotomizing axons projecting to the lumbar muscle and knee in rats. *Eur. Spine,* 12, 576-580.

O'Neill, J., Brock, C., Olesen, A. E., Andresen, T., Nilsson, M., Dickenson, A. H. (2012). Unravelling the mystery of capsaicin: a tool to understand and treat pain. *Pharmacol. Rev.,* 64, 939-971.

Pecze, L., Blum, W., Schwaller, B. (2013). Mechanism of capsaicin receptor TRPV1-mediated toxicity in pain-sensing neurons focusing on the effects of Na+/Ca2+ fluxes and the Ca2+-binding protein calretinin. *Mol. Cell Res.,* 1833, 1680-1691.

Perry, M. J., Lawson, S. N., Robertson, J. (1991). Neurofilament immunoreactivity in populations of rat primary afferent neurons: a quantitative study of phosphorylated and non-phosphorylated subunits. *Neurocytol.,* 20, 746-758.

Perl, E. R. (1996) Cutaneous polymodal receptors: Characteristics and plasticity. *Prog. Brain Res.,* 113, 21-37.

Piper, A. S. and Docherty, R. J. (2000). One-way cross-desensitization between P2X purinoceptors and vanilloid receptors in adult rat dorsal root ganglion neurons. *Physiology,* 15, 685-696.

Porseva, V. V. (2013). Characteristics of capsaicin-sensitive afferent neurons of spinal nerve ganglia. *Cell and Tissue Biology,* 7, 149-153.

Porseva, V. V., Shilkin, V. V., Korzina, M. B., Korobkin, A. A., Maslyukov, P. M. (2012a). Changes in TRPV1-immunoreactive neurons in spinal nerve sensory ganglion neurons in rats on exposure to capsaicin. *Neurosci. Behav. Physiol.,* 42, 770-774.

Porseva, V. V., Strelkov, A. A., Shilkin, V. V., Masliukov, P. M. (2012b). Age-related changes in sensory neurons containing calcitonin gene-related peptide under conditions of afferentation deficit in rats. *Ontogenez,* 43, 405-12.

Porseva, V. V., Shilkin, V. V., Korzina, M. B., Korobkin, A. A., Maslyukov, P. M. (2013a). Substance P-immunopositive neurons in rat spinal nerve. *Neurosci. Behav. Physiol.,* 43, 150-152.

Porseva, V. V., Shilkin, V. V., Korzina, M. B., Smirnova, V. P., Maslyukov, P. M. (2013b). Developmental changes in $NF200^{+}$ neurons in sensory ganglia at different segmental levels on chemical deafferentation. *Neurosci. Behav. Physiol.,* 43, 602-606.

Poyner, D. R., Sexton, P. M., Marshall, I., Smith, D. M., Quirion, R., Born, W., Muff, R., Fischer, J. A., Foord, S. M. (2002). International Union of

Pharmacology. XXXII. The mammalian calcitonin gene-related peptides, adrenomedullin, amylin, and calcitonin receptors. *Pharmacol. Rev.,* 54, 233-246.

Price, T. J., Flores, C. M. (2007). Critical evaluation of the colocalization between calcitonin gene-related peptide, substance P, transient receptor potential vanilloid subfamily type 1 immunoreactivities, and isolectin B4 binding in primary afferent neurons of the rat and mouse. *Pain,* 8, 263-72.

Reyes-Escogido, M., Gonzalez-Mondragon, E. G. and Vazquez-Tzompantzi, E. (2011). Chemical and pharmacological aspects of capsaicin. *Molecules,* 16, 1253-1270.

Rosenfeld, M. G., Mermod, J. J., Amaraa, S. G., Swanson, L. W., Swchenko, P. E., Rivier, J., Vale, W. W., Evans, R. M. (1983). Production of a novel neuropeptide encode by the calcitonin gene via tissue-specitfic RNA processing. *Nature,* 304, 129-135.

Russo, D., Clavenzani, P., Mazzoni, M., Chiocchetti, R., Di Guardo, G., Lalatta-Costerbosa, G. (2010). Immunohistochemical characterization of TH13-L2 spinal ganglia neurons in sheep (Ovis aries). *Microsc. Res. Тесн.,* 73, 128-139.

Saffroy, M., Beaujouan, J. C., Torrens, Y., et al. (1987). Localization of tachykinin binding sites (NK1, NK2, NK3 ligands) in the rat brain. *Peptides*, 9, 227-241.

Schicho, R., Skofitsch, G., Donnerer, J. (1999). Regenerative effect of human recombinant NGF on capsaicin-lesioned sensory neurons in the adult rat. *Brain Res.,* 815, 60-69.

Schicho, R., Donnerer, J., Liebmann, I., Lippe, I. T. (2005). Nociceptive transmitter release in the dorsal spinal cord by capsaicin-sensitive fibers after noxious gastric stimulation. *Brain Res.,* 1039, 108-115.

Shigemoto, R., Yokota, Y., Tsuchida, K., Nakanishi, S. (1990). Cloning and expression of a rat neuromedin K receptor cDNA. *Biol. Chem.,* 265, 623-628.

Shadiack, A. M., Sun, Y. and Zigmond, R. E. (2001). Nerve growth factor antiserum induces axotomy-like changes in neuropeptide expression in intact sympathetic and sensory neurons. *Neurosci.,* 21, 363-371.

Simone, D. A., Nolano, M., Johnson, T., Wendelschafer-Crabb, G., and Kennedy, W. R. (1998). Intradermal injection of capsaicin in humans produces degeneration and subsequent reinnervation of epidermal nerve fibers: correlation with sensory function. *Neurosci.,* 18, 8947-8959.

Snider, W. D., McMahon, S. B. (1998). Tackling pain at the source: new ideas about nociceptors. *Neuron,* 4, 629-632.

Suzuki, R., Morcuende, S., Webber, M., Hunt, S. P., Dickenson, A. H. (2002). Superficial NK1-expressing neurons control spinal excitability through activation of descending pathways. *Nat. Neurosci.,* 5, 1319-1326.

Szallasi, A. and Blumberg, P. M. (1999). Vanilloid (capsaicin) receptors and mechanisms. *Pharmacol. Rev.,* 51, 159-211.

Szallasi, A., Cortright, D. N., Blum, C. A., Eid, S. R. (2007). The vanilloid receptor TRPV1: 10 years from channel cloning to antagonist proof-of-concept. *Nat. Rev. Drug Discov.*, 6, 357-372.

Tajti, J., Uddman, R., Moller, S., et al. (1999). Messenger molecules and receptor mRNA in the human trigeminal ganglion. *Auton. Nerv. Syst.,* 76, 176-183.

Tonra, J. R., Mendell, L. M. (1998). Effects of postnatal anti-NGF on the development of CGRP-IR neurons in the dorsal root ganglion. *Comp. Neurol.,* 392, 489-498.

Tominaga, M., Tominaga, T. (2005). Structure and function of TRPV1. *Pflugers Arch.,* 451, 143-150.

Torrealba, F. (1992). Calcitonin gene-related peptide immunoreactivity in the nucleus of the tractus solitarius and the cardiac receptors of the cat originates from peripheral afferents. *Neurosci.,* 47, 165-173.

Venkatachalam, K. and Montell, C. (2007). TRP channels. *Annu. Rev. Biochem.,* 76, 387-417.

Van Rossum, D., Hanisch, U. K., Quirion, R. (1997). Neuroanatomical localization, pharmacological characterization and functions of CGRP, related peptides and their receptors. *Neurosci. Biobehav. Rev.,* 21, 649-678.

Wiley, R. G., Kline, R. L. III and Vierck, C. J. (2007). Anti-nociceptive effects of selectively destroying substance P receptor-expressing dorsal horn neurons using [Sar9,Met(O2)11] substance P-saporin: behavioral and anatomical analyses. *Neurosci.,* 146, 1333-1345.

Willis, W. D., Coggeshall, R. E. (2004). *Sensory Mechanisms of the Spinal Cord. 3rd ed.* New York, New York, US: Kluwer Academic/Plenum.

Winter, J., Evison, C. J., O'Brien, C., Benowitz, L., Lindsay, R. M., Mulderry, P., Woolf, C. (1992). Neurotoxic damage evokes regenerative responses from adult rat sensory neurones. *Neurosci. Lett.,* 146, 48-52.

Woodbury, C. J., Zwick, M., Wang, S., Lawson, J. J., Caterina M. J., Koltzenburg, M., Albers, K. M., Koerber, H. R., and Davis, B. M. (2004). Nociceptors lacking TRPV1 and TRPV2 have normal heat responses. *Neurosci.,* 24, 6410-6415.

Xu, P., Van Slambrouck, C., Berti-Mattera, L., Hall, A. K. (2005). Activin induces tactile allodynia and increases calcitonin gene-related peptide after peripheral inflammation. *Neurosci.,* 25, 9227-9235.

Yasuchika, A., Seiji, O., Kazuhisa, T., Hidetoshi, I., Hideo, D., Tomoyuki, O., Tomoko, S., Hideshige, M. (2005). Expression and co-expression of VR1, CGRP, and IB4-binding glycoprotein in dorsal root ganglion neurons in rats: differences between the disc afferents and the cutaneous afferents. *Spine,* 30, 1496-1500.

Zubrzycka, M., Janecka, A. (2000). Substance P: Transmitter of nociception. *Endocrine Regulation,* 34, 195-201.

In: Capsaicin
Editor: Bruce Gilliam
ISBN: 978-1-61470-433-1

Chapter 7

Capsaicin-Loaded Nanoemulsions as a Functional-Ingredient Delivery System and Its Biological Evaluation

Chong-Tai Kim[1,*], Yangha Kim[2], Ae-Jin Choi[3] and Jin-Soo Maeng[1]

[1]Research Group of Convergence Technology, Korea Food Research Institute, Seongnam-si, Gyeonggi-do, Republic of Korea
[2]Department of Nutritional Science and Food Management, Ewha Womans University, Seoul, Republic of Korea
[3]National Institute of Horticultural and Herbal Science, Suwon, Gyeonggi-do, Republic of Korea

Abstract

Capsaicin is widely consumed as a spice and food additive, and is also utilized in medical applications. Capsaicin (N-[4-hydroxy-3-methoxyphenyl] methyl-8-methyl-non-6-enamide) and dihydrocapsaicin (N-[4-hydroxy-3-methoxy-phenyl]methyl-8-methyl-non-6-nonanamide) are two members of the family of naturally occurring capsaicinoids, which make up the pungent components of hot pepper. Capsaicin has been obtained in the oleoresin capsicum form through the industrial extraction of the dried ripe fruits of capsicums, and it contains a complex

* Corresponding author: Email: ctkim@kfri.re.kr.

mixture of essential oils, waxes, colored materials, and several capsaicinoids. Capsaicin is one of the carotenoids, such as capsanthin, cryptoxanthin, carotene and zeaxanthin, which contribute to the yellow to red colours of many foods. However, there are several limitations for its use in food systems, such as lipophilicity and the structure of carotenoids, which affect their distribution in the cellular systems *in vivo*, and high melting point, making them crystalline at food storage and body temperatures. One of the promising nanotechnologies is the nanometer-sized emulsion delivery system, which offers the potential to improve the solubility and bioavailability of many bioactive ingredients such as carotenoids, polyunsaturated fatty acids, phytosterols, antioxidants, vitamin, minerals, and other natural compounds. Bioactive ingredients can be encapsulated in nanometer-sized structures, which function as the transport systems. These nanometer-sized structures have sizes ranging from 20–500 nm, and provide the necessary physical properties, such as solubility, stability, and bioavailability. A nanopolymeric system can be applied to the formulation of nanoemulsions through self-assembly emulsification for the development of a bioactive ingredient delivery system. The incorporation of alginate and chitosan to form multi-layer nanoemulsions can be expected to improve the stability of nanoemulsions, and this system holds promise for use in the production of functional foods containing bioactive ingredients. The nanoemulsions of capsaicin which incorporated with alginate (AN) were found to reduced the body weight and adipose tissue mass, whereas capsaicin extract (CE) did not. CE and AN reduced the levels of mRNA of adipogenic genes, including peroxisome proliferator-activated receptor-γ, sterol regulatory element-binding protein-1c, and fatty acid-binding protein in white adipose tissue. The mRNA levels of genes related to β-oxidation or thermogenesis, including PPAR-α, palmitoyltransferase-1α, and uncoupling protein-2, were increased by supplementation with CE and AN relative to the high fat control group. In particular, the expression of PPAR-α, CPT-1α, and UCP2 was well as the activity of AMPK were significantly increased in the AN group compared to those in the OC group.

1. Introduction

Red pepper (*Capsicum annuum* L.) has been cultivated and used in many countries including Brazil, China, Honduras, Guatemala, India, Japan, Mexico, Mombassa, Pakistan, the Republic of Korea, Uganda, Thailand, Zanzibar, and the United States. The total yearly production of red pepper is about 20 million tons, representing roughly 30 billion dollars in the world market. Among market areas, the dried red pepper spices market is estimated to account for 6 billion dollars as a food ingredient, and to2 billion dollars as spices [1]. Red pepper is used in many different dishes depending on the types of food in countries, such as Mexican, Italian, and Indian. Capsaicinoids are major compounds in hot peppers, which is mostly consists of capsaicin and dihydrocapsaicin, making up 80-90% of the total capsaicinoids [2]. Several techniques have been employed for the extraction of capsaicin from red peppers, such as Soxhlet [3], ultrasound-assisted extraction [4], supercritical fluid extraction [5], and microwave-assisted extraction [6]. It can be extracted from fresh green chili peppers at a 400% greater yield than other methods [7]. Capsaicin is produced in the form of oleoresin capsicum in an industrial extraction of the dried ripe fruits of capsicums, and contains a complex mixture of essential oils, waxes, colored materials, and several capsaicinoids.

Capsaicin (the main pungent ingredient in hot pepper) is known as a useful functional ingredient because it can prevent oxidative stress-mediated chronic diseases such as cancer, cardiovascular disease, hypertension, stroke, and neurodegenerative disorders, due to its antioxidant activity in the absence of toxicity [8]. It is widely consumed as a spice and food additive, and is also utilized in medical applications [9-11]. Capsaicin and dihydrocapsaicin are insoluble in water and fat-soluble compounds. In rats, the absorption of the capsaicinoids was reported to be about 60-80% [12]. Nanoemulsion technology has been successfully applied to [roduce nano-ingredients for functional foods, such as with capsaicin and other spice oleoresins [13, 14]. However, most of the spice oleoresins from red pepper, garlic, ginger, onion, etc., exhibit sensitivity to light, heat and oxygen, have low water solubility, and have short storage lives if not stored properly. Oil-in-water (O/W) emulsions have been used as vehicles for the delivery of lipophilic bioactives in the food and drug industry. However, food emulsions are thermodynamically unstable systems and will eventually breakdown due to the increase in interfacial area after emulsification. The bioactive ingredients can be encapsulated in nanometer-sized structures, which function as transport systems. These nanometer-sized structures have sizes ranging from 20–500 nm and provide the necessary physical properties, such as solubility, stability, and bioavailability [15]. This chapter details current research on the preparation, characterization and biological evaluation of food nanoemulsion and its applications of capsaicin-loaded nanoemulsions for the use of a functional-ingredient delivery system for noodle-type food products.

2. Extraction and Purification of Capsaicin

Extraction of capsaicin from green chili peppers was attempted in order to increase the yield and the purity of the isolated capsaicin, for which the method was in compliance with preliminary standards issued by the National Cancer Institute [7]. This method is summarized as a flow diagram in Figure 1. When compared with extractions of dried red peppers, using the above described procedures, the use of fresh green peppers as the starting material resulted in increase of the yields by about 6 fold. After extraction and purification of capsaicin, about 1.12% yield was achived. When the same procedure was applied to extract dried hot red peppers, a yield of only about 0.562% was obtained. Due to the properties of capsaicin, distillation must be conducted under a vacuum of at least 4×10^{-5} mmHg to prevent degradation of the compound.

Ethanol and acetonitrile are better solvents for the extraction of capsaicin from fresh samples, while acetone was better for dried pepper parts. The yields of dihydrocapsaicin from fresh samples are also typically higher with acetonitrile and ethanol [16]. It has been reported that the polarity of solvents affects the extraction efficiencies of capsaicinoids [17]. Capsaicinoid extraction from capsicum fruits and oleoresins using solvents such as acetone, chloroform, methanol, acidified methanol and acetonitrile demonstrated acetone to result in the highest yields from dehydrated ground capsicum [18]. However, when producing capsaicin extract for use in the food industry, safety regulations are necessary in order to avoid the production of capsaicin extract containing toxic organic solvents for the protection of health.

3. Capsaicin-Loaded Nanoemulsions

3.1. Preparation of Nanoemulsion from Capsaicin

Capsaicin is insoluble in water. Its poor water solubility (400 μg/ml), may also be responsible for its poor bioavailability, as dissolution is a rate-limiting factor in the intestinal absorption of poorly water-soluble drugs [19]. Various edible oils and surfactants were tested herein to determine the solubility through use of the shake or sonication method. Briefly, an excess amount of capsaicin was added to vials containing the selected vehicles, including edible oil, surfactant or co-surfactant. After sealing, the mixture was homogenized, shaken, centrifuged and filtered. Finally, the samples were evaluated by macroscopy for the selection of a suitable surfactant, co-surfactant and edible oil. In order to prepare the pre-mixtures, it was necessary to examine varying compositions of capsaicin, surfactant, and co-surfactant based on the weight of each component. In this step, a ternary phase diagram of the surfactant, co-surfactant and edible oil can be plotted, each of them, representing an apex of the triangle. The samples which maintained a transparent phase after a certain period were considered to be nanoemulsions. The formation of nanoemulsions in ternary mixtures varies with the compositions of surfactant, co-surfactant and edible oil. The ternary phase diagram of the capsaicin/surfactant/water system is presented in Figure 2. The phase boundary was determined by characterizing the particle size of the nanoemulsion. The white area represents the one-phase nanoemulsion, with the transparent region ranging below 100 nm. The grey area is the multiple, cloudy region of the emulsion, ranging between 100 and 400 nm. Finally, the dark grey area is the region of phase separation, which occurred in both the capsaicin and surfactant corners of the phase diagram.

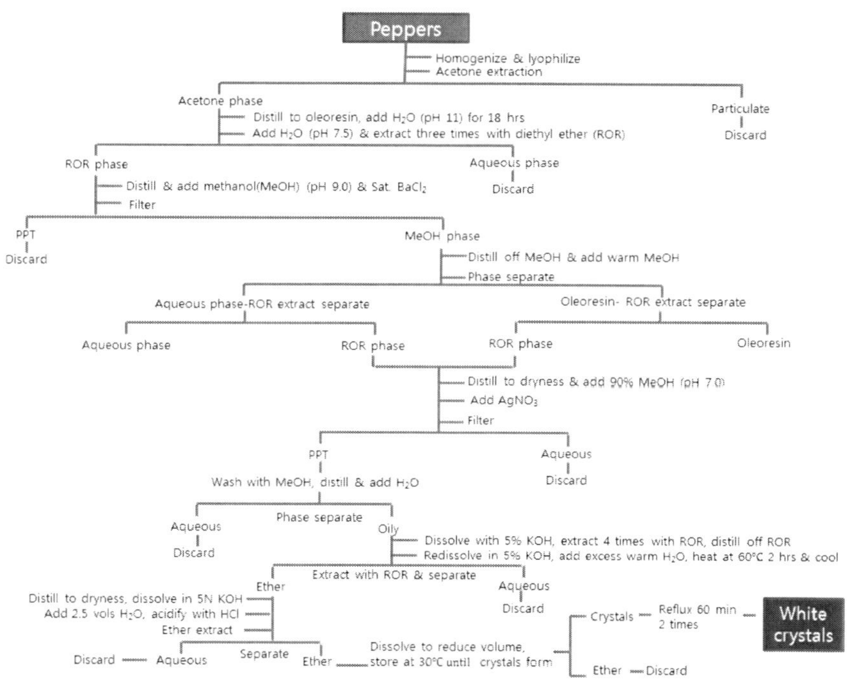

Figure 1. Extraction process of capsaicin from fresh pepper.

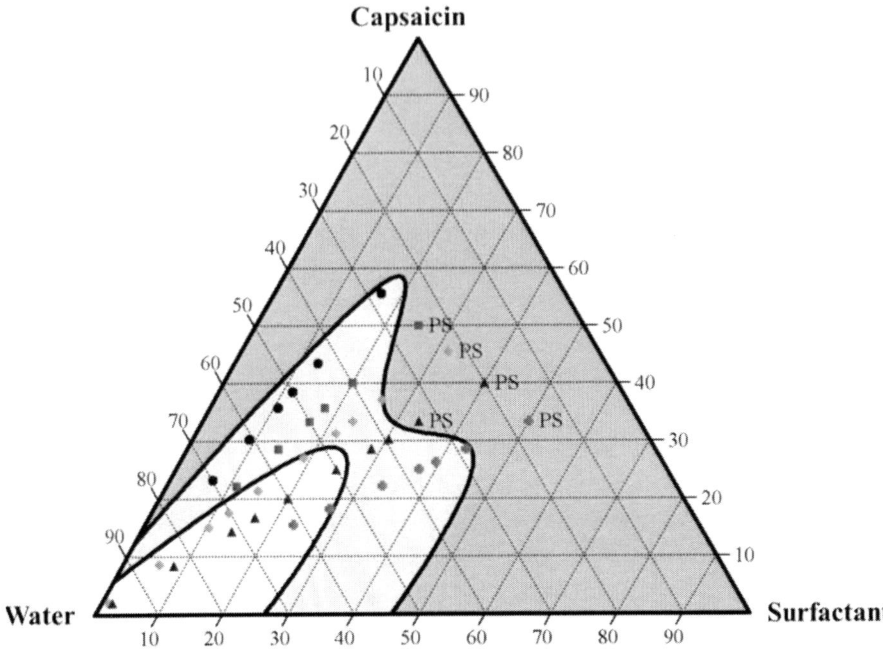

Figure 2. Ternary phase diagram of capsaicin-loaded nanoemulsions with the capsaicin/surfactant/water system.

3.2. Characterization of Capsaicin-Loaded Nanoemulsions

The factors which must be taken into account for quality control in the preparation of nanoemulsions are stability and bioavailability. Many of the most important properties of nanoemulsion-based food products are determined by the size of the particles, which affect characteristics such as shelf life, appearance, texture and flavor. Therefore, control, prediction, measurement, and report of the droplet size in nanoemulsions are the determining parameters. The size distribution of capsaicin-loaded nanoemulsions obtained herein is illustrated in Figure 3. Investigations on their physical stability showed no significant changes in the particle size distributions of the three samples, however, the volume of smaller particles increased with the increasing ration of surfactant to capsaicin without the occurrence of tailing, which might be due to coalescence in the capsaicin nanoemulsions. This means that the ratio of capsaicin and surfactant must be a considerable parameter in the formulation of nanoemulsions.

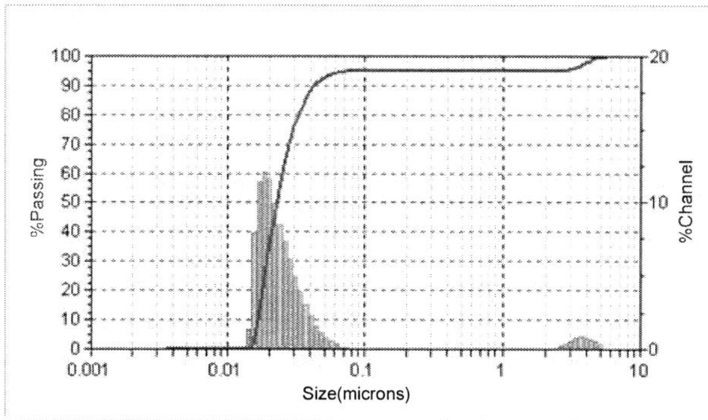

Capsaicin : Surfactant = 1 : 3

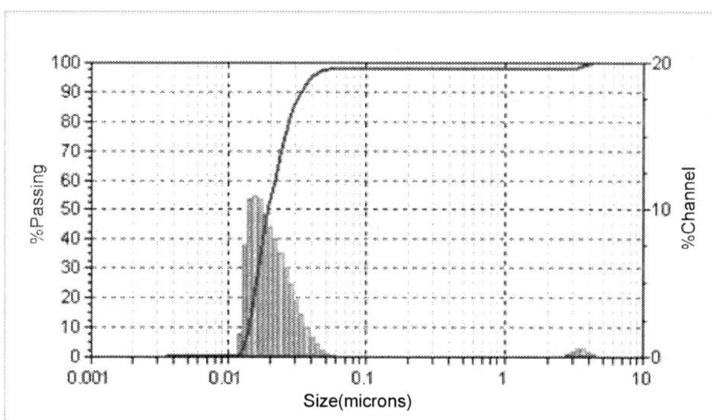

Capsaicin : Surfactant = 1 : 4

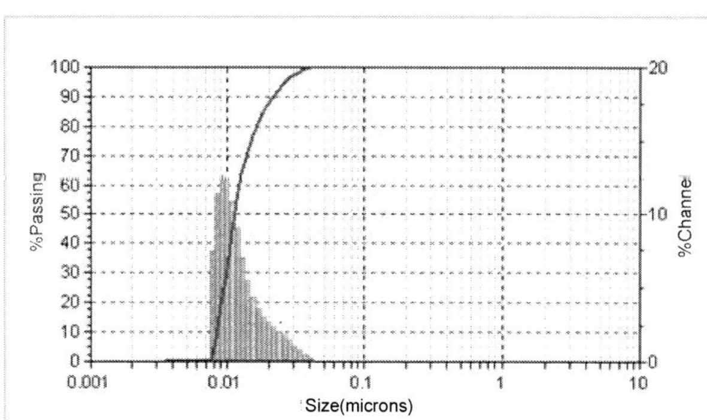

Capsaicin : Surfactant = 1 : 5

Figure 3. Particle size distribution of capsaicin-loaded nanoemulsions prepared according to the ratios of capsaicin and surfactant.

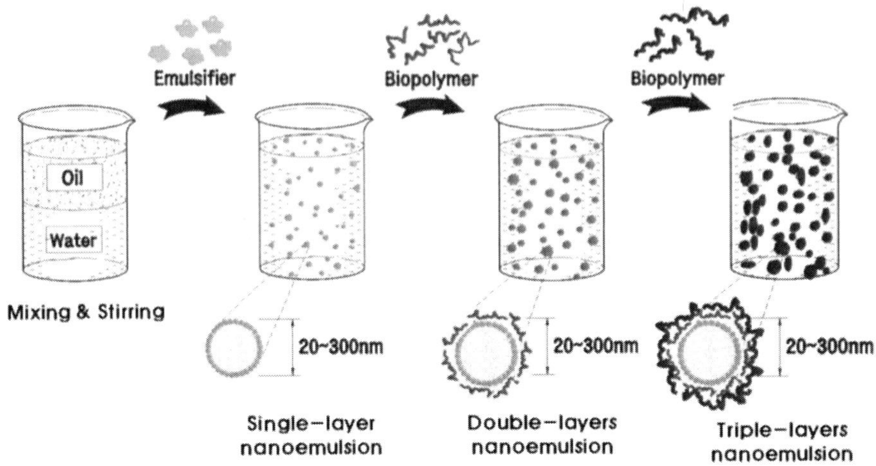

Figure 4. Schematic representation of the formation of single and multilayer capsaicin-loaded nanoemulsions.

Emulsion particles may have electrical charge depending on the type of surface-active molecules present and the pH of the aqueous phase. The droplets in most nanoemulsions have an appreciable electrical charge, and therefore, electrostatic interactions may play an important role in determining their overall stability and physicochemical properties. The zeta potential is often a better representation of the electrical characteristics of an emulsion droplet because it inherently accounts for the adsorption of any charged counter ions [20]. The zeta potential represents the electric charge between the shear plane of the final external layer and the bulk solution, and this parameter has been used to estimate the property of dispersoids. This factor is influenced by the composition of the nanoemulsions and the electrical phenomenon.

3.3. Multilayer Nanoemulsions of Capsaicin

As food delivery systems, simple solutions or nanoemulsions may be unstable or degradable due to changes in the nature of the emulsion structure and in environmental conditions such as temperature, light, and pH. However, it is possible to develop nanoparticles which are novel delivery systems with the use of food-grade biopolymers such as polysaccharides (alginate, chitosan, pectin, gum arabic) and proteins (casein, gelatin, whey) by promoting self-association or aggregation of single biopolymers, or by inducing phase separation in mixed biopolymer systems [21]. A schematic representation of the formation of multilayer emulsions and the procedure for the preparation of nanoemulsions based on alginate/chitosan are shown in Figure 4. A single-layer nanoemulsion was formed by stirring oil and aqueous phases together in the presence of a water-soluble emulsifier. A double-layer nanoemulsion was then formed by coating the single-layer nanoemulsion with biopolymer. Finally, triple-layer nanoemulsion was obtained by adding polyelectrolytes to an emulsion containing oppositely charged biopolymer, such as anion-charged alginate and cation-charged chitosan.

The single-, double-, and triple-layer nanoemulsions fabricated using 0.05% chitosan and 0.05% alginate solutions appeared transparent and flowed easily, as an oil-in-water (O/W) nanoemulsion, indicating that the nanoemulsions contained particle sizes below 100 nm (Figure 5). Fabrication of the double-layer nanoemulsion using the 0.05% chitosan solution as the aqueous phase resulted in the formation of particles with a minimum size of 10 nm. The particle size of the double-layer nanoemulsion based on the 0.05%, 0.15%, 0.2%, and 0.6% chitosan solutions significantly increased when subjected to heat treatment at 70 °C ($p<0.01$). In addition, the viscosity of the nanoemulsions increased with increase in the concentration of chitosan. In contrast, no significant changes were observed for nanoemulsions fabricated using the 0.05% chitosan solution as the aqueous phase after heat treatment ($p<0.05$). The triple-layer nanoemulsions manufactured using 0.05% chitosan and 0.05% alginate solutions had the highest incorporation efficiency, 68.00±0.34%, which significantly decreased when the concentration of chitosan was higher than 0.05% ($p<0.001$). Therefore, the chitosan and alginate concentrations which resulted in the highest incorporation efficiency and storage stability were determined to be 0.05%. The high incorporation efficiency of the triple-layer nanoemulsions occurred because the polycationic nature of chitosan leads to a strong interaction with the negatively charged alginate. When alginate was dropped into chitosan solution, the electrostatic interaction between the carboxylic groups of alginate and the amine groups of chitosan resulted in the formation of a membrane [22].

Herein, four nanoemulsions based on different biopolymers were prepared: SN, single-layer nanoemulsion based on aqueous phase; AN, double-layer nanoemulsion based on alginate; CN, double-layer nanoemulsion based on chitosan; CAN, triple-layer nanoemulsion based on chitosan/alginate. Statistically significant differences in the zeta potential were observed among SN, AN, CN, and CAN ($p<0.01$, Table 1). The zeta potential of SN was –14.23±1.96 mV, while that of CN was 26.33±12.68 mV at 25 °C due to the positive charge of the chitosan molecules coated on the surface of CN. In contrast, the zeta potential of AN was –31.70±0.36 mV, due to the negative charge of the alginate molecules. The zeta potential of CAN was –8.97±0.43 mV, which was lower than SN. These findings suggest that the carboxyl group of alginate interacts with the amine group of chitosan, reducing the number of positive charges exposed on the surface of the nanoemulsion. The results confirmed the formation of an alginate and chitosan layer the nanoemulsion surface. The absolute value of the zeta potential can be used as a measure of the stability of the nanoemulsions; therefore, CN and AN were comparatively more stable than SN. However, due to the coexistence of anions and cations by bonding with the deposited electrical ions, the stability of CAN was lower than SN, CN and AN.

The morphological properties of the biopolymer nanoemulsions fabricated using chitosan and alginate were imaged by TEM (Figure 6). The TEM images indicated the presence of a droplet-type nanoemulsion structure, which existed in the single-, double-, and triple-layer forms. The hydrophilic portions of the droplets were stained black, while the hydrophobic components were unstained and ranged in size from 25 to 70 nm. SN, the single-layer nanoemulsion solution, was not uniform in shape and was partially bound to large or small emulsion drops. The CN, AN, and CAN double- or triple-layer nanoemulsions were trapped within the charged biopolymers chitosan, alginate, or both, respectively. CN, AN, and CAN showed a more compact and clear particle structure, which may be due to flocculation and coalescence caused by phase separation or stabilization of the nanoemulsions.

Table 1. Polydispersity, zeta potential and pH of capsaicin-loaded nanoemulsions

	Polydispersity index	Zeta potential (mV)	pH
SN	0.47±0.01^{b}	-14.2±1.96^{bc}	5.24±0.12^{a}
CN	0.63±0.12^{a}	26.3±12.7^{ab}	4.49±0.12^{b}
AN	0.39±0.08^{b}	-31.7±0.36^{a}	5.65±0.25^{a}
CAN	0.66±0.01^{a}	-8.97±0.43^{c}	5.41±0.43^{a}
F-value	10.02*	8.05*	11.23*

Results were expressed as the mean±SEM (n=3). Means with different letters in the same column differ significantly by Duncan's multiple range test ($p<0.05$).

SN, single-layer nanoemulsion based on aqueous phase; CN, double-layer nanoemulsion based on chitosan; AN, double-layer nanoemulsion based on alginate; CAN, triple-layer nanoemulsion based on chitosan/alginate. *Significantly different at the $P<0.01$ level.

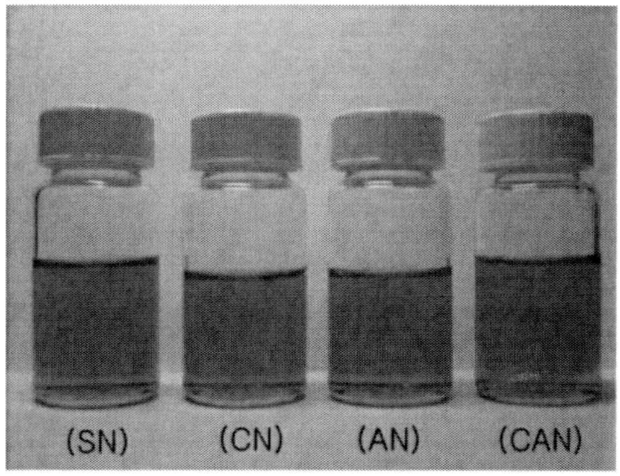

Figure 5. Difference in appearance of the capsaicin-loaded nanoemulsions. SN, single-layer nanoemulsion based on aqueous phase; CN, double-layer nanoemulsion fabricated with chitosan; AN, double-layer nanoemulsion fabricated with alginate; CAN, triple-layer nanoemulsion fabricated with chitosan/alginate.

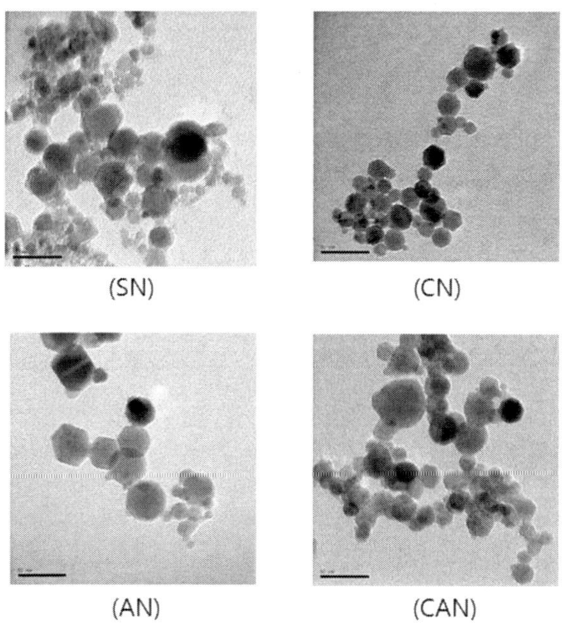

Figure 6. TEM of single-, double and triple-layer capsaicin-loaded nanoemulsions fabricated with chitosan and alginate. SN, Single-layer nanoemulsion based on aqueous phase; CN, double-layer nanoemulsion fabricated with chitosan; AN, double-layer nanoemulsion fabricated with alginate; CAN, triple-layer nanoemulsion fabricated with chitosan/alginate.

4. Stability of Capsaicin-Loaded Nanoemulsions

Oil-in-water (O/W) emulsions have been used as vehicles for the delivery of lipophilic bioactives in the food and drug industry. However, food emulsions are thermodynamically unstable systems, and will eventually break down due to increase in the interfacial area after emulsification. Consequently, the physical instability, such as creaming, flocculation, coalescence, phase conversion and Ostwald ripening can occur. Recently, novel fabrication methods for the production of O/W emulsions with long-term stability have been investigated at the Korea Food Research Institute. The bioactives are encapsulated in nanometer-sized structures, which are assembled with both core materials and biopolymers, to result in transparent or translucent systems mostly covering the size range of 20-200 nm, called nanoemulsions. The stability of single layer nanoemulsions of capsaicin showed dependence on the particle size and zeta potential during long term storage over 2 years (Figure 7). The particle size of capsaicin nanoemulsions was relatively maintained in the range of 35-50 nm, indicating that no insoluble aggregates were formed. Although the zeta potential decreased slightly from -52 to -30 mV after 2 years of storage, this still represents a stable range. This result can be attributed to the fact that the capsaicin-loaded nanoemulsions are highly charged, so there is strong electrostatic repulsion between them. In addition, the magnitude of the electrical charge on the capsaicin-loaded nanoemulsions was relatively small, which suggested that these molecular complexes would form either soluble complexes or coacervates.

5. Applications of Capsaicin-Loaded Nanoemulsions as a Food Ingredient

Recently, the instant noodle market has been experiencing rapid growth in Korea and other Asian countries. This situation is due to two main factors: first, the booming of Korean culture including food, entertainment and tourism, and second, the new market trend in formulating food products with functional actives that can deliver unique health benefits.

Red pepper has been extensively used as a seasoning ingredient in traditional Korean cuisine due to its red color, unique sweet taste and pungency. Especially, the dried red pepper powder or flakes have been used as seasoning ingredients together with garlic, ginger and onion for the soup base in instant fried noodles. However, these types of seasoning ingredients are difficult and inconvenient to handle when added to instant fried noodles due to their powdery properties, which might csause dust to be raised and scattered. The use of capsaicin as a food nano-ingredient that can offer several advantages over traditional delivery systems for functional ingredients, which include increases in its solubility, stability and bioavailability.

Noodles containing capsaicin were prepared with formulated capsaicin-loaded nanoemulsions, and control noodles were prepared without capsaicin (Figure 8). After cooking the noodles, the capsaicinoid content in noodles prepared with SN, AN and CN were decreased by 34.9, 27.7 and 19.3%, respectively, while the control was reduced by 40.2% (Table 2). The slight decrease of capsaicinoids in the noodles prepared with capsaicin-loaded nanoemulsions was due to the insolubility of amylose-lipid complexes. The difference of capsaicinoid content in noodles between SN, AN and CN may also be due to flocculation and coalescence caused by phase separation or stabilization of the nanoemulsions.

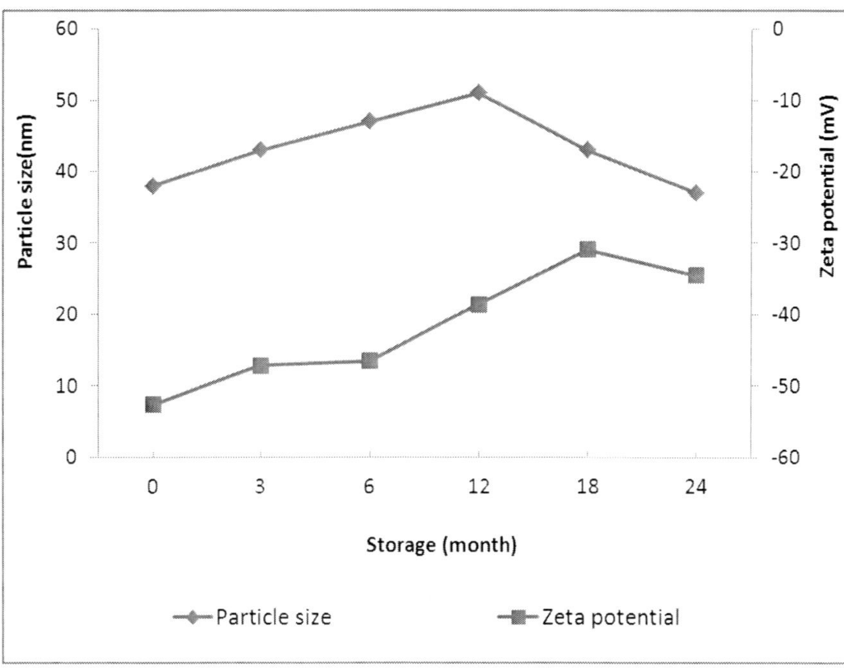

Figure 7. Long term stability of capsaicin-loaded nanoemulsions depending on the particle size and zeta potential.

Table 2. Capsaicinoids content in noodles formulated with capsaicin-loaded nanoemulsions

Noodles	Capsaicinoids (μg/g)		Capsaicinoids loss (%)
	Wet noodle	Cooked noodle	
CON	2.39±0.04	1.43±0.05	40.2±0.55
SN	2.20±0.03	1.43±0.04	34.9±0.74
AN	2.24±0.02	1.62±0.03	27.7±0.67
CN	2.23±0.02	1.80±0.03	19.3±0.52

CON, noodles based on premixture of capsaicin emulsion; SN, noodles based on capsaicin-loaded single layer nanoemulsion; AN, noodles prepared with capsaicin-loaded double layer nanoemulsion based on alginate; CN, noodles prepared with capsaicin-loaded double layer based on chitosan.

(a) (b)

Figure 8. Photograph of noodles prepared with capsaicin-loaded nanoemulsion. A, Wet noodles (Control); B, Noodles prepared with the capsaicin-loaded double layer nanoemulsion fabricated with alginate (AN).

6. Biological Evaluation of Capsaicin-Loaded Nanoemulsions

The most abundant and well-known bioactive compound of chili peppers obtained from plants of the genus capsicum is capsaicin. Capsaicin, along with dihydrocapsaicin, forms a naturally occurring chemical group called capsaicinoids, which are responsible for the pungent flavor of peppers [23].

Besides their culinary effects, the bioactive effects of capsaicinoids have also gained interest for health outcomes. Clinical human trials have suggested that capsaicinoids may have anti-cancer [24], anti-inflammatory [25] and antioxidant properties [26], as well as pain-relief action [27]. In addition, there is accumulating evidence that capsaicinoids may induce weight reduction. Accumulating evidence from interventional studies has demonstrated that capsaicinoids result in the increase of energy expenditure [28], thermogenesis and lipid oxidation [29, 30], the reduction of body fat accumulation [30] and the regulation of energy intake [31]. In 3T3-L1 adipocytes, the extraction of hot pepper was shown to lead to a decrease in adipogenesis by reducing the activity of glycerol-3-phosphate dehydrogenase (GPDH) and expression of CCAAT-enhancer-binding protein (C/EBP)-α and C/EBP-β mRNA [32]. Recent study also demonstrated the anti-obesity activity of capsaicin by inhibiting adipogenesis in 3T3-L1 cells via activation of transient receptor potential vanilloid 1 receptor (TRPV1), inducing brown-like phenotype [33].

In spite of the beneficial effects of capsaicin on human health, the practical use and application are limited due to its low solubility and poor bioavailability [13]. Capsaicin is produced in the form of capsaicin extract (CE) in the industry. CE is an organic solvent extract of dried ripe red pepper and contains acids, esters, alcohols, aldehydes, ketones, and carotenoid pigments, as well as capsaicin analogs [35]. Nanometer-sized emulsion delivery systems have been regarded as a promising biotechnology for the improvement of physiological performance, solubility and bioavailability [36]. These improvements in the capsaicin-loaded nanoemulsions fabricated with alginate (AN) have been expected to improve health more than CE, however, studies have not been executed to investigating the molecular and intracellular mechanisms by which AN improves health outcomes compared to CE.

To date, only one animal study directly investigated the effects of CE and AN on obesity in rats with high fat (HF) diet-induced obesity [37]. A 14 week HF diet challenge with AN was found to alter the body weight and fat accumulation in adipose tissue without changing food intake. AN supplementation significantly decreased body weight by 10.2%, compared to the CE group (Table 3). In addition, rats fed HF plus AN had significant reduction of the white adipose tissue mass (WAT) by 17%, whereas no difference were observed for the CE group (Table 3). As shown in Figure 9, not only WAT mass, but also adipocyte size was distinctly smaller in the group fed AN compared to CE. There were no significant changes in liver weight. The mRNA levels of adipogenic genes including peroxisome proliferator-activated receptor (PPAR)-γ, sterol regulatory element-binding protein-1c, and fatty acid-binding protein in WAT were significantly decreased in both the CE and AN supplemented groups compared to the HF control group, as shown in Figure 10A. However, AMP-activated protein kinase (AMPK) activity (Figure 10C) and mRNA expression of genes involved in β-oxidation or thermogenesis, including PPAR-α, palmitoyltransferase-1α, and uncoupling protein-2 (Figure 10B), were significantly higher in the AN group than the CE group. Therefore, AN showed favorable effects on obesity in obese rats fed HF diet, being even more effective than CE. Further investigation is needed to determine whether AN directly helps with weight management in obese human subjects using long-term clinical trials with different dosages incorporated with the diet.

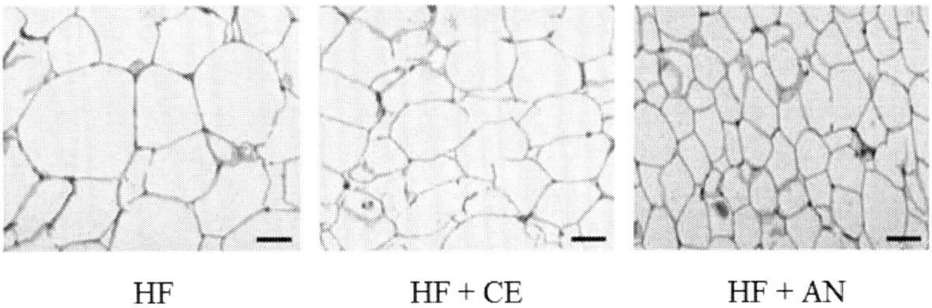

Figure 9. Effects of AN on adipocyte size. Representative histological sections of epididymal adipose tissue (hematoxylin and eosin stain; magnification 100×; scale bar =100 μm). HF, high-fat; AN, capsaicin-loaded nanoemulsion double nanoemulsion fabricated with alginate; CE, capsaicin extract.

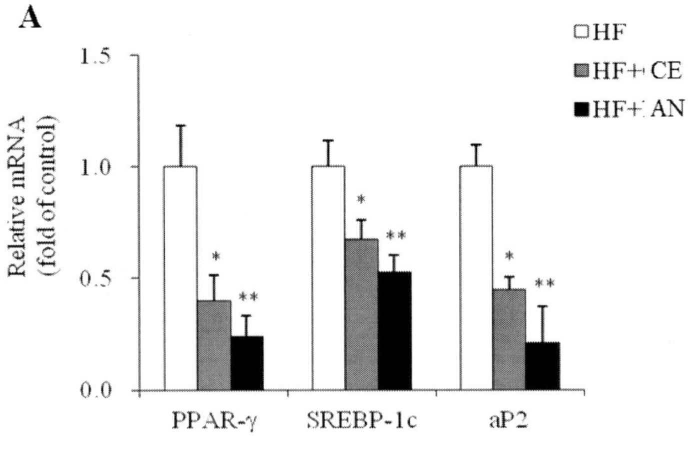

Figure 10. Continued on next page.

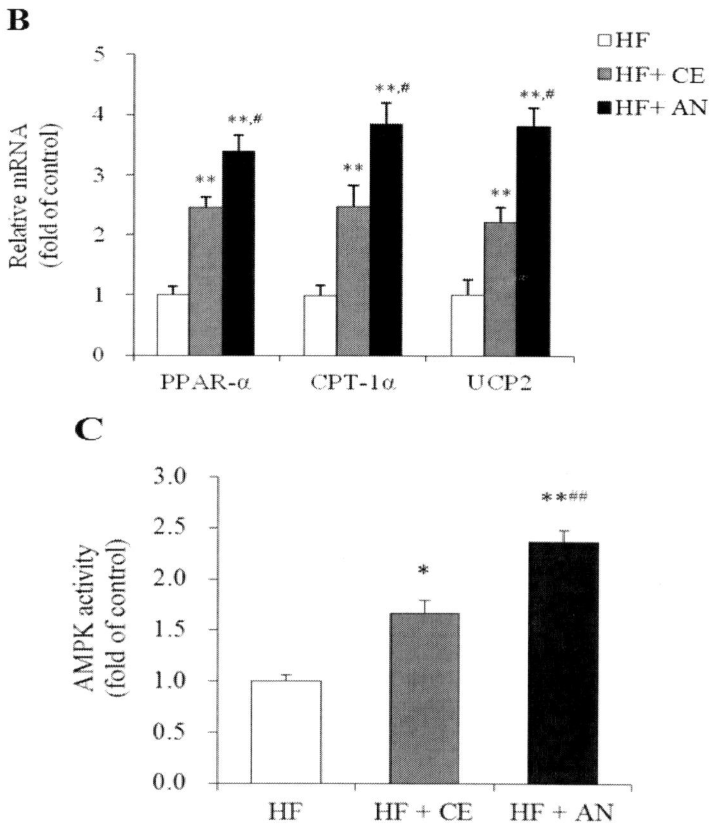

Figure 10. Effects of CE and AN on the mRNA expression of genes involved in energy metabolism and AMPK activity in WAT. Relative changes in mRNA levels of genes involved in (A) adipogenesis, including PPAR-γ, SREBP-1c, and aP2, in (B) β-oxidation and thermogenesis, including PPAR-α, CPT-1α, and UCP2, and in (C) AMPK activity. Total RNA samples were isolated from different treatment groups, and analyzed by real-time PCR, and normalized to β-actin. AMPK activity was measured using an AMPK kinase kit and normalized to the protein content. Data were expressed as fold increase of the HF group (n=7). *$P<0.05$; **$P<0.01$ versus HF group; #$P<0.05$ versus HF + CE group. AMPK, AMP-activated protein kinase; aP2, fatty acid-binding protein 2; CPT, carnitine palmitoyltransferase; HF, high-fat; AN, capsaicin-loaded nanoemulsion double nanoemulsion fabricated with alginate; CE, capsaicin extract; PPAR, peroxisome proliferator-activated receptor; SREBP, sterol regulatory element-binding protein; UCP2, uncoupling protein-2; WAT, white adipose tissue.

Table 3. Effects of capsaicin and capsaicin-loaded nanoemulsion on metabolic characteristics

	HF	HF + CE	HF + AN
Final body weight (g)	699.9±15.4	662.1±8.1	628.6±22.3*
Food Intake (g/day)	25.3±0.4	23.8±0.2	23.7±0.7
Liver weight (g/100 g body weight)	2.9±0.2	2.9±0.1	3.1±0.1
Adipose tissue weight (g/100 g body weight)	3.34±0.07	3.58±0.18	3.1±0.13*

HF, high fat diet; AN, capsaicin-loaded nanoemulsion based on alginate; CE, capsaicin extract; Asterisk indicates statistical difference compared to HF+OC ($p<0.05$).

CONCLUSION

There are technical challenges that need to be overcome to impart the functionality of lipophilic bioactives lacking in solubility, stability, and bioavailability to the human body. This results in the unsuccessful enhancement of quality and commercialization in the food industry. To overcome this problem, it is important to attempt to understand the requirements for solubilization of the oil phase for the development of a stable emulsion by the use of food grade components such as emulsifiers and stabilizers to produce food oil-soluble supplements and nutraceuticals. Food nanomaterials, such as nanoemulsions, nanoparticles, nanocomposites and nanostructured ingredients, may generally be prepared by using nanotechnology, including nanoemulsification, association colloid and nanostructured multiple or multilayer emulsification. The bioactives are encapsulated in nanometer-sized structures, which are assembled with both core materials and biopolymers, in transparent or translucent systems mostly covering the size range of 20-200 nm, called nanoemulsions. In addition, it is possible to develop multilayer emulsions providing a wide range of different functionalities by using different combinations of surfactants and biopolymers for the production of novel food delivery systems. Capsaicin-loaded nanoemulsions may especially create a market demand for new products using the effect of enhanced solubility, stability and bioavailability of emulsion products or already known extracts. The capsaicin-loaded nanoemulsions showed favorable effects on obesity in obese rats fed with high fat diets, and were more effective than regular capsaicin extracts. Further investigation is needed to determine whether capsaicin-loaded nanoemulsions directly help with weight management in obese human subjects using long-term clinical trials with different dosages incorporated in the diet. Consequently, there is the need for certain applications to employ novel and alternative types of food delivery systems that can be produced in the food and beverage industries.

REFERENCES

[1] Park, J. B., Lee, M. A. & Hwang, I. K. (2012). Research report on "Research group of export product for Korean red pepper", Gyeonggi-do, Republic of Korea, Korea Food Research Institute.

[2] Constant, H. L. & Cordell, G. A. (1997). *Studies in Natural Product Chemistry*, 2nd ed. Elsevier Science Publishers, USA.

[3] Korel, F., Bagdatlioglu, N., Balaban, M. O. & Hisil, Y. (2002). Ground red peppers: Capsaicinoids content, Scoville Scores, and discrimination by an electronic nose. *J. Agric. Food Chem.*, *50* (11), 3257–3261.

[4] Barbero, G. F., Liazid, A., Palma, M. & Barroso, C. G. (2008). Ultrasound-assisted extraction of capsaicinoids from peppers. *Talanta*, *75* (5), 1332–1337.

[5] Duarte, C. M. M., Crew, M., Casimiro, T. & Aguiar, R. A. & Nunesda, P. M. (2002). Phase equilibrium for capsaicin+ water+ethanol+ supercritical carbon dioxide. *J. Supercrit. Fluids*, *22* (2), 87–92.

[6] Barbero, G. F., Palma, M. & Barroso, C. G. (2006). Determination of capsaicinoids in peppers by microwave-assisted extractionhigh-performance liquid chromatography with fluorescence detection. *Anal. Chim. Acta*, *25* (2), 227–233.

[7] Sass, N. L., Rounsavil, M. & Combs, H. (1977). A high-yield method for the extraction and purification of capsaicin. *J. Agric. Food Chem.*, *25* (6), 1419–1420.

[8] Jorge, M. T., Abraham, G. C. & Enrique, R. C. (1999). Antimicrobial properties of alkamides present in flavoring plants traditionally used in Mesoamerica: Affinin and capsaicin. *J. Ethnopharmacol.*, *64*, 241-248.

[9] Kopec, S. E., Debellis, R. J. & Irwin, R. S. (2002). Chemical analysis of freshly prepared and stored capsaicin solutions: Implications for tussigenic challenges. *Pulm. Pharmacol. Ther.*, *15*, 529-534.

[10] Lee, K.W., Lee, H. J., Surh, Y. J. & Lee, C. Y. (2003). Vitamin C and cancer chemoprevention: Reappraisal. *Am. J. Clin. Nutr.*, *78*, 1074-1078.

[11] Reilly, C. A., Ehlhardt, W.J., Jackson, D. A., Kulanthaivel, P., Mutlib, A. E., Espina, R. J., Moody, D. E., Crouch, D. J. & Yost, G. S. (2003). Metabolism of capsaicin by cytochrom P450 produces novel dehydrogenated metabolites and decrease cytotoxicity to lung and liver cells. *Chem. Res. Toxicol.*, *16*, 336-339.

[12] Kawada, T. & Iwai, K. (1985). In Vivo and In Vitro metabolism of dihydrocapsaicin, a pungent principle of hot peppers, in rats. *Agric. Biol. Chem.*, *49*, 441-448.

[13] Choi, A. J., Kim, C. J., Cho, Y. J., Hwang, J. K. & Kim, C. T. (2009). Effects of surfactants on the formation and stability of capsaicin-loaded nanoemulsions. *Food Sci. Biotechnol.*, *18*(5), 1161-1172.

[14] Choi, A. J., Kim, C. J., Cho, Y. J., Hwang, J. K. & Kim, C. T. (2011). Characterization of capsaicin-loaded nanoemulsions stabilized with alginate and chitosan by self-assembly. *Food Bioprocess Technol.*, *4*, 1119-1126.

[15] Finsher, J. H. (1968). Particle size of drugs and its relationship to absorption and activity. *J. Pharm. Sci.*, *57*, 1825–1835.

[16] Chinn, M. S., Sharma-Shivappa, R. R. & Cotter, J. L. (2011). Solvent extraction and quantification of capsaicinoids from Capsicum chinense. *Food and Bioproducts Processing*, *89*, 340-345.

[17] Peusch, M., Seitz, E. M., Petz, M., Muller, A. & Anklam, E. (1997). Extraction of capsaicinoids from chilies (Gapsicum fruteseens L.) and paprika (Gapsieum annum L.) using supereritieal fluids and organic solvents. *Eur. Food Res. Technol. (Z. Lebensm. Unters. Forsch. A)*, *204*, 351-355.

[18] Attuquayefio, V. K. & Buckle, K. A. (1987). Rapid sample preparation method for HPLC analysis of capsaicinoids in capsicum fruits and oleoresins. *J. Agrie. Food Chem.*, *35*, 777-779.

[19] Nielloud, F. & Marti-Mestres, G. (2000). *Pharmaceutical emulsions and suspensions*. New York: Marcel Dekker.

[20] McClements, D. J., Decker, E. A. & Weiss, J. (2007). Emulsion-based delivery systems for lipophilic bioactive components. *Journal of Food Science*, *72*(8), R109–R124.

[21] Weiss, Weiss, J., Takhistov, P. & Mcclements, J. (2006). Functional materials in food nanotechnology. *Journal of Food Science*, *71*(9), 107–116.

[22] GåserØd, O., SmiderØd, O. & Skjåk-Braek, G. (1998). Microcapsules of alginate-chitosan-I A quantitative study of the interaction between alginate and chitosan. *Biomaterials*, *19*, 1815–1825.

[23] Mueller-Seitz, E., Hiepler, C. & Petz, M. (2008). Chili pepper fruits: content and pattern of capsaicinoids in single fruits of different ages. *J. Agric. Food Chem.*, *56*(24), 12114–12121.

[24] Yang, Z. H., Wang, X. H., Wang, H.P., Hu, L. Q., Zheng, X. M. & Li, S. W. (2010). Capsaicin mediates cell death in bladder cancer T24 cells through reactive oxygen species production and mitochondrial depolarization. *Urology*, *75*(3), 735–741.

[25] Choi, S. E., Kim, T. H., Yi, S. A., Hwang, Y. C., Hwang, W. S., Choe, S. J., Han, S. J., Kim, H. J., Kim, D. J., Kang, Y. & Lee, K. W. (2011). Capsaicin attenuates palmitate-induced expression of macrophage inflammatory protein 1 and interleukin 8 by increasing palmitate oxidation and reducing c-Jun activation in THP-1 (human acute monocytic leukemia cell) cells. *Nutr. Res.*, *31*(6), 468–478.

[26] Henning, S. M., Zhang, Y., Seeram, N. P., Lee, R. P., Wang, P., Bowerman, S. & Heber, D. (2011). Antioxidant capacity and phytochemical content of herbs and spices in dry, fresh and blended herb paste form. *Int. J. Food. Sci. Nutr.*, *62*(3), 219–225.

[27] Fraenkel, L., Bogardus, Jr. S. T., Concato, J. & Wittink, D. R. (2004). Treatment options in knee osteoarthritis: the patient's perspective. *Arch. Intern. Med.*, *164*(12), 1299–1304.

[28] Yoshioka, M., Lim, K., Kikuzato, S., Kiyonaga, A., Tanaka, H., Shindo, M. & Suzuki, M. (1995). Effects of red-pepper diet on the energy metabolism in men. *J. Nutr. Sci. Vitaminol.*, (Tokyo), *41*(6), 647-56.

[29] Lee, T. A., Li, Z., Zerlin, A. & Heber, D. (2010). Effects of dihydrocapsiate on adaptive and diet-induced thermogenesis with a high protein very low calorie diet. A randomized control trial. *Nutr. Metab.* (Lond), 7, 78.

[30] Snitker, S., Fujishima, Y., Shen, H., Ott, S., Pi-Sunyer, X., Furuhata, Y., Sato, H. & Takahashi, M. (2009). Effects of novel capsinoid treatment on fatness and energy metabolism in humans: possible pharmacogenetic implications. *Am. J. Clin. Nutr.*, *89*(1), 45-50.

[31] Westerterp-Plantenga, M. S., Smeets, A. & Lejeune, M. P. (2005). Sensory and gastrointestinal satiety effects of capsaicin on food intake. *Int. J. Obes.*, (Lond). *29*(6), 682–688.

[32] Jeon, G., Choi, Y., Lee, S. M., Kim, Y., Jeong, H. S. & Lee, J. (2010). Anti-obesity activity of methanol extract from hot pepper (Capsicum annuum L.) seeds in 3T3-L1 adipocyte. *Food Sci. Biotechnol.*, *19*(4), 1123–1127.

[33] Baboota, R. K., Singh, D. P., Sarma, S. M., Kaur, J., Sandhir, R., Boparai, R. K., Kondepudi, K. K. & Bishnoi, M. (2014). Capsaicin Induces "Brite" Phenotype in Differentiating 3T3-L1 Preadipocytes. *PLoS One*, *9*(7), e103093.

[34] Olajos, E. J., Salem, H. (2001). Riot control agents: pharmacology, toxicology, biochemistry and chemistry. *J. Appl. Toxicol.*, *21*(5), 355–391.

[35] Huang, Q., Yu, H. & Ru, Q. (2010). Bioavailability and delivery of nutraceuticals using nanotechnology. *J. Food. Sci.*, *75*(1), R50–R57.

[36] Kim, J. Y., Lee, M. S., Jung, S., Joo, H., Kim, C. T., Kim, I. H., Seo, S., Oh, S. & Kim, Y. (2014). Anti-obesity efficacy of nanoemulsion oleoresin capsicum in obese rats fed a high-fat diet. *Int. J. Nanomedicine.*, *9*, 301-310.

In: Capsaicin
Editor: Bruce Gilliam

ISBN: 978-1-61470-433-1

Chapter 8

INFLUENCE OF CAPSAICIN ON LIPID PEROXIDATION AND ANTIOXIDANT STATUS DURING ACETAMINOPHEN INDUCED EXPERIMENTAL HEPATOTOXICITY

P. Anandakumar* and T. Devaki
Department of Biochemistry, University of Madras,
Chennai, India

ABSTRACT

The protective role of capsaicin on oxidative stress produced during acetaminophen induced toxic hepatitis was studied in rats.

Administration of acetaminophen (800 mg/kg body weight intraperitoneally) to rats caused a significant increase in oxidative stress indicated by abnormal changes in lipid peroxidation, enzymic antioxidants (superoxide dismutase, catalase, glutathione peroxidase, glutathione reductase and glutathione-S-transferase) and non-enzymic antioxidants (reduced glutathione, vitamin C, vitamin E and vitamin A).

Treatment with capsaicin (10 mg/kg body weight intraperitoneally) reverted all the above abnormal changes and restored cellular normalcy

* Dr. P. Anandakumar, Flat No.G2, Lovely apartments, Door No.32/12, Amman Koil Street, Srinivasa Nagar Post, New Perungalathur, Chennai – 600 063. TAMILNADU, INDIA. Tel (Mobile): +91 9840694276, E-mail: bioanand77@gmail.com.

suggesting the modulatory efficacy of capsaicin in curbing the oxidative stress produced during acetaminophen induced hepatitis.

Keywords: Capsaicin, acetaminophen, hepatitis, lipid peroxidation, antioxidants

INTRODUCTION

There is ample evidence of increased toxicity in our food, air and water. In addition, chronic illnesses often are connected to the production of toxic chemicals that inflame and cause oxidative stress.

Liver is affected by different hepatotoxins including alcohol, aflatoxin, heavy metals and drugs. Out of this acetaminophen is one of the well established as suitable experimental model of liver toxicity (Werowatganon et al., 2014).

Accumulating evidence suggests that free radical mediated oxidative stress plays a major role in varieties of toxicities. (Devaki et al., 2009). High levels and/or inadequate removal of reactive oxygen species may cause severe metabolic imbalance and oxidative damage to biological macromolecules resulting in lipid peroxidation and depletion of cellular antioxidants (Sathivel et al., 2008).

Varieties of studies have shown that dietary intake of foods rich in fruits and vegetables decreased the occurrence of hepatitis (Wang et al., 1999; Gyamfi et al., 1999).

Pepper fruits (Capsicum annuum L.) are important vegetables used as vegetable foods and as spices with well-known antioxidant properties (Materska Malgorzata and Perucka Irena, 2005).

Capsaicin (CAP) (trans-8-methyl-N-vannilyl-6-nonemide) is the major pungent principle found in hot red and chilli peppers of the plant genus Capsicum that has long been used as spices, food additives and drugs (Cordell and Araiyo, 1993).

Different studies on CAP have also revealed the antioxidant activity of CAP in several dimensions (Rice Evan, 1997; Lugman and Rizvi, 2006; De and Ghosh, 1993, Anandakumar et al., 2008a).

In our present study the ability of CAP in modulating the oxidative stress produced during acetaminophen induced rat hepatitis was studied using lipid peroxidation and antioxidant parameters.

MATERIALS AND METHODS

Materials

Acetaminophen and CAP were purchased from M/s.Sigma chemicals, St. Louis, USA. All other chemicals were of analytical grade procured from M/s. SRL Chemicals Pvt. Ltd., Mumbai.

Animals

Adult male albino rats of Wistar strain weighing around 120 to 150 g obtained from Tamil Nadu Veterinary and Animal Sciences University (TANUVAS), Madhavaram, Chennai, India were used in this study. The animals were housed under conditions of controlled temperature (26 ±2°C) with 12-hr day/night cycle. They were fed standard rat/mice pellet diet (M/s. Hindustan Lever Ltd., Mumbai) under the trade name Amrut rat/mice feed and were given access to water *ad libitum.*

Experimental Design

Experimental animals were divided into four groups of six rats each as follows.

Group I: control animals received olive oil throughout the course of the experiment

Group II: animals were treated with a single dose of acetaminophen (800 mg/kg body weight) intraperitoneally (Balaji Raghavendran et al., 2005)

Group III: animals received CAP alone (10 mg/kg body weight dissolved in olive oil) intraperitoneally to assess the cytotoxicity (if any) induced by CAP

Group IV: animals received acetaminophen (as in Group II) along with CAP (10 mg/kg body weight dissolved in olive oil) intraperitoneally. CAP pre treatment was started 4 weeks prior to the administration of acetaminophen

At the end of the experimental period, animals were sacrificed by cervical decapitation under ether anesthesia and liver was excised immediately and washed with ice-cold saline. A 10% homogenate of the washed tissue (liver) was prepared in 0.01 M phosphate buffer (pH 7.4). The homogenate was centrifuged at a speed of 12,000 x g for 30 min in a refrigerated high-speed centrifuge at 4°C. Blood was also collected and the serum was separated for other estimations. The following analysis was carried out in the supernatant.

Biochemical Analysis

Protein estimation was carried out (Lowry et al., 1951). LPO was assayed by the method in which the malondialdehyde (MDA) released served as the index (Ohkawa et al., 1979). Superoxide dismutase (SOD) and catalase (CAT) were assayed (Marklund and Marklund, 1974; Sinha, 1974). The activities of glutathione peroxidase (GPx), glutathione reductase (GR) and glutathione-S-transferase (GST) were also was determined. (Rotruck et al., 1973; Beutler, 1968; Habig et al., 1974). Reduced glutathione (GSH), vitamin E, vitamin C and vitamin A were assayed (Moron et al., 1979; Desai, 1984; Omaye et al., 1971; Bayfield and Cole., 1980)

Data Analysis

All data were expressed as mean ± S.D for six rats. The results were computed statistically (SPSS Software Package) using one-way ANOVA. Post-hoc testing was performed for inter comparisons using the LSD. ($P < 0.05$) was considered significant.

Results

Effect of Acetaminophen and Capsaicin on Lipid Peroxidation in Rat Liver

Figure 1 shows the effect of acetaminophen and CAP on liver lipid peroxides (MDA) levels. Acetaminophen induced group II rats showed significantly ($P < 0.05$) increased levels of MDA in liver compared to control rats (Group I). Pre treatment with CAP (Group IV) markedly ($P < 0.05$) decreased the levels of MDA in liver when compared with Group II animals.

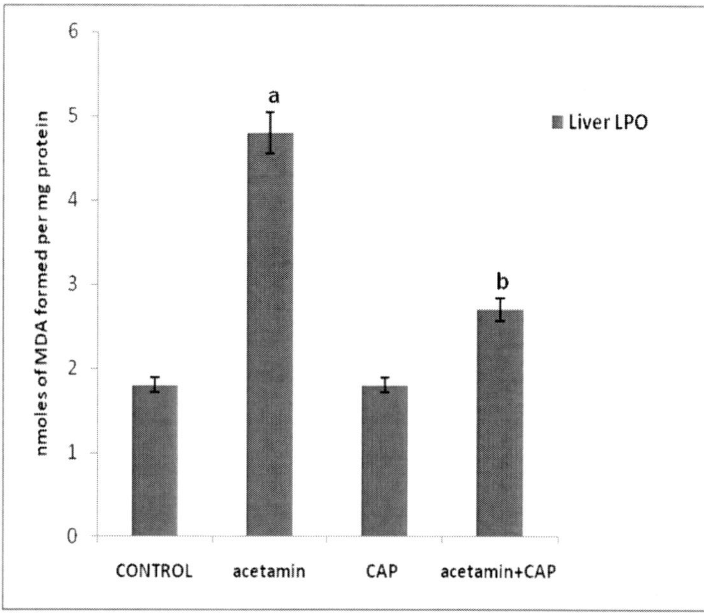

Each value is expressed as mean ± S.D for six rats in each group. Statistical significance at $P<0.05$, as compared with [a]Group I, [b]Group II.

Figure 1. Effect of acetaminophen and capsaicin on lipid peroxidation in rat liver.

Effect of Acetaminophen and Capsaicin on Enzymic Antioxidants in Rat Liver

Table 2 depicts the effect of acetaminophen and CAP on liver enzymic antioxidant status. Significant ($P < 0.05$) reduction in the activities of enzymic antioxidants SOD, CAT, GPx, GR and GST were observed in the acetaminophen administered group (Group II). CAP pretreatment resulted in a free radical quenching effect and thereby significantly ($P < 0.05$) reinstated the enzymic antioxidant activities to near normalcy in Group IV animals.

Effect of Acetaminophen and Capsaicin on Non-Enzymic Antioxidants in Rat Liver

Table 3 shows the levels of non-enzymic antioxidants GSH, vitamin C, vitamin E and vitamin A in control and experimental group of animals.

Acetaminophen treatment markedly ($P < 0.05$) reduced the levels of GSH, vitamin C, vitamin E and vitamin A in Group II animals that were significantly ($P < 0.05$) restored to near normalcy on CAP pretreatment in Group IV animals.

Table 1. Effect of acetaminophen and capsaicin on enzymic antioxidants in rat liver

	Group I (Control)	Group II (acetaminophen)	Group III (CAP)	Group IV (acetaminophen + CAP)
SOD	7.44 ± 0.07	3.48 ± 0.04[a]	7.26 ± 0.09	5.88 ± 0.04[b]
CAT	75.13 ± 1.30	55.03 ± 1.31[a]	74.59 ± 1.23	69.66 ± 1.27[b]
GPx	107.33 ± 1.85	59.71 ± 1.08[a]	106.06 ± 2.53	88.68 ± 1.77[b]
GR	5.39 ± 0.43	1.98 ± 0.13[a]	5.37 ± 0.43	4.08 ± 0.08[b]
GST	239.22 ± 6.20	148.36 ± 3.81[a]	238.64 ± 7.25	216.51 ± 7.06[b]

Each value is expressed as mean ± S.D for six rats in each group.

SOD – units/mg protein. (One unit is equal to the amount of enzyme required to inhibit autoxidation of pyrogallol by 50%); CAT - µg of H_2O_2 consumed/min/mg protein; GPx - µg of reduced GSH utilized/min/mg protein; GR - nmoles of NADPH oxidized/min/mg protein; GST - nmoles of CDNB - GSH conjugate formed/min/mg protein;

Statistical significance at $P<0.05$, as compared with [a]Group I, [b]Group II.

Table 2. Effect of acetaminophen and capsaicin on non-enzymic antioxidants in rat liver

	Group I (Control)	Group II (acetaminophen)	Group III (CAP)	Group IV (acetaminophen + CAP)
GSH	7.61 ± 0.37	2.96 ± 0.10[a]	7.55 ± 0.34	6.88 ± 0.32[b]
Vitamin C	5.32 ± 0.23	2.48 ± 0.18[a]	5.31 ± 0.15	3.88 ± 0.5[b]
Vitamin E	1.88 ± 0.25	0.57 ± 0.04[a]	1.82 ± 0.10	1.39 ± 0.19[b]
Vitamin A	0.58 ± 0.05	0.20 ± 0.0 [a]	0.54 ± 0.05	0.39 ± 0.04[b]

Each value is expressed as mean ± S.D for six rats in each group.

GSH – µg/mg protein; vitamin C – µg/mg protein; vitamin E – µg/mg protein; vitamin A – µg/mg protein.

Statistical significance at $P<0.05$, as compared with [a]Group I, [b]Group II.

Discussion

Any pathological state that leads to increased production and/or ineffective scavenging of reactive oxygen species may play a crucial role in determining tissue injury. Free radicals are continuously produced during different drug toxicities. Thus generated free radicals target macromolecules, such as DNA, protein and lipids, with the ultimate progression of lipid peroxidation (Sriram et al., 2008). Increased LPO products in cells can result in cellular dehydration, whole cell deformity and cell death. Thus LPO is considered as the prime rationale that instigates liver injury (Sathivel et al., 2008a). Our results were in consistent with the previous findings as we observed increased liver LPO in group II acetaminophen induced rats.

The major mechanism accountable for alkaloid substances in combating the ROS induced oxidative damage is due to the antioxidant effect of these compounds (Anandakumar et al., 2008a). CAP is one such alkaloid whose potential antioxidant role has been reported (Surh et al., 1995). Previous reports confirm that the radical scavenging site of CAP is the C7-benzyl carbon (Kogure et al., 2002). The above reports were substantiated by our present study, where pretreatment with CAP reduced liver LPO induced by acetaminophen in group IV animals suggesting its antioxidant efficacy.

To prevent oxidative damage in the cell, a variety of antioxidants scavenge free radicals. The primary defense against oxidative stress in the tissue rests with antioxidants. SOD, an important antioxidant enzyme protects cells against the toxic effects of superoxide anion (Ravikumar et al., 2005b).

CAT catalyses the direct degradation of H_2O_2 to water. It protects the cellular constituents against oxidative damage (Balaji Raghavendran et al., 2005). GPx and the cellular NADPH generating mechanisms together form a system for removing the hydroperoxides from the cell (Balaji Raghavendran et al., 2005). GST is a group of multifunctional proteins that perform tasks ranging from catalyzing the detoxification of electrophilic compounds to protection against peroxidative damage (Anandakumar et al., 2008b). GR plays a major role in regenerating GSH from GSSG, thus maintaining the balance between the redox couple (Kamaraj et al., 2007).

The decreased activities of these antiperoxidative enzymes during acetaminophen administration in the present study are compatible with earlier findings (Balaji Raghavendran et al., 2005). The decreased activity of these enzymatic antioxidants may be due to the accumulation of H_2O_2 which in turn causes the inhibition of these enzymes. The reduced activities of these enzymes were normalised upon pretreatment with CAP.

GSH, a chief endogenous antioxidant system is found in particularly high concentration in liver and is known to have key functions in protective processes (Pushpavalli et al., 2010). The protective role of GSH against cellular lipid peroxidation has been well recognized (Sheriff and Devaki, 2013). Vitamin C a major antioxidant acts in tissues, involving ROS in aqueous phase and it has been reported that the tissue concentration of vitamin C is a good indicator of oxidative stress (El-Beshbishy, 2008). Vitamin E is a principal lipid soluble antioxidant in cell membranes that protects critical cellular structures against oxidative damage (El-Beshbishy, 2008). The concentration of vitamin E has been inversely correlated to LPO (Sheriff and Devaki, 2013). Vitamin A as an antioxidant is involved in counteracting free radicals and is known to help in the repair of damaged tissues (Serbecic et al., 2005). Vitamin A at pharmacological doses is reported to ameliorate the oxidative stress mediated membrane lipid peroxidation and membranes enriched with vitamin A are protected against oxidative stress *in vivo* and exhibit resistance to LPO induced *in vitro* (Britton et al., 2000). Vitamin C, E and A levels were found to be reduced in acetaminophen administered groups (Group II) suggesting an increase in LPO in these animals. The levels of these vitamins in rats challenged with acetaminophen were almost normalized upon CAP pretreatment. Our results were in consistent with Kempaiah et al. (Kempaiah et al., 2004) who reported that CAP was effective against oxidative stress by countering the depleted antioxidant molecules and antioxidant enzymes.

Thus the results of our present study exemplify the role of CAP as a strong antioxidant compound that curbed oxidative stress and in turn paved way for its beneficial effect against acetaminophen induced hepatitis in rats.

ACKNOWLEDGMENTS

First author Anandakumar P wishes to thank Mr.Pandi, Mrs. Padmavathi, Mrs.Kalaivani and Mrs. Prabhavathy for their moral support and encouragement throughout the period of study.

REFERENCES

Anandakumar, P., Kamaraj S., Jagan S., Ramakrishnan G., Vinodhkumar R., Devaki T. (2008a). Capsaicin modulates pulmonary antioxidant defense system during benzo(a)pyrene-induced lung cancer in Swiss albino mice. *Phytotherapy Research,* 22: 529–533.

Anandakumar, P., Kamaraj, S., Jagan, S., Ramakrishnan, G., Vinodhkumar R., Devaki T. (2008b). Stabilization of pulmonary mitochondrial enzyme system by capsaicin during benzo(a)pyrene induced experimental lung cancer. *Biomedicine and Pharmacotherapy,* 62: 390–394.

Balaji Raghavendran, H.R., Sathivel, A., Devaki, T. (2005). Antioxidant effect of Sargassum polycystum (Phaeophyceae) against acetaminophen induced changes in hepatic mitochondrial enzymes during toxic hepatitis. *Chemosphere,* 61(2):276-281.

Bayfield, R.F., Cole, E.R. (1980). Colorimetric estimation of vitamin A with trichloroacetic acid. *Methods in Enzymology,*67:189.

Beutler, E. (1968). Active transport of glutathione disulfide from erythrocytes. *In:* Larson A, Orrenius S, Holmgren A, Mannerwik B, editors. *Functions of glutathione:Biochemical, physiological, toxicological and clinical aspects.* New York: Raven Press; p. 65.

Cordell, G.A., Araiyo, O.E. (1993). Capsaicin identification, nomenclature and pharmacotherapy. *Annals of Pharmacotherapy,* 27:330.

De, A.K., Ghosh, J.J. (1993). Capsaicin pretreatment protects free radical induced rat lung damage on exposure to gaseous chemical lung irritants. *Phytotherapy Research,*7:87.

Desai, I.D. (1984). Vitamin E analysis methods for animal tissues. *Methods in Enzymology,* 105:138.

Devaki, T., Sathivel, A., BalajiRaghavendran, H.R. (2009). Stabilization of mitochondrial and microsomal function by polysaccharide of Ulva lactuca on D-Galactosamine induced hepatitis in rats. *Chemico Biological Interactions,* 177(2):83-88.

El-Beshbishy, H.A. (2008). Aqueous garlic extract attenuates hepatitis and oxidative stress induced by galactosamine/lipoploysaccharide in rats. *Phytotherapy Research,* 22(10):1372-1379.

Gyamfi, M.A., Yonamine, M., Aniya, Y. (1999). Free radical scavenging action of medicinal herbs from Ghana.thonningia sanguine on experimentally-induced liver injuries. *General Pharmacology*, 32: 661-667.

Habig, W.H., Pabst, M.J., Jacob, W.B. (1974). Glutathione-S-transferase. The first enzymatic step in mercapturic acid formation. *Journal of Biological Chemistry,* 249:7130.

Han, K.H., Hashimoto, N., Shimada, K., Sekikawa, M., Noda, T., Yamauchi, H., Hashimoto, M., Chiji, H., Topping, D.L., Fukushima, M. (2006).

Hepatoprotective effects of purple potato extract against D-galactosamine-induced liver injury in rats. *Bioscience, Biotechnology and Biochemistry,* 70(6):1432-1437.

Kamaraj, S., Vinodhkumar, R., Anandakumar, P., Jagan, S., Ramakrishnan, G. (2007). The effects of quercetin on antioxidant status and tumor markers in the lung and serum of mice treated with benzo(a)pyrene. *Biological Pharmaceutical Bulletin,* 30:2268–2273.

Kempaiah, P., Srinivasan, K. (2004). Influence of dietary curcumin, capsaicin and garlic on the antioxidant state of red blood cells and the liver in high-fat fed rats. *Annals of Nutritional Metabolism,*48:314.

Kogure Kentaro., Goto Satoru., Nishimura Miki., Yasumoto Mina., Abe Kazutoyo., Ohiwa Che. (2002).. Mechanism of potent antiperoxidative effect of capsaicin. *Biochima et Biophysics Acta*, 1573:84.

Lowry, O.H., Rosenbrough, N.J., Farr, A.L., Randall R.J. (1951). Protein measurement with the Folin's phenol reagent. *Journal of Biological Chemistry,* 193:265.

Lugman, S., Rizvi, S.I. (2006). Protection of lipid peroxidation and carbonyl formation in proteins by capsaicin in human erythrocytes subjected to oxidative stress. *Phytotherapy Research,* 20:303.

Marklund, S., Marklund, G.(1974). Involvement of superoxide anion radical in the autooxidation of pyrogallol and a convenient assay for superoxide dismutase. *European Journal of Biochemistry,* 47:469.

Materska Malgorzata., Perucka Irena. (2005). Antioxidant activity of the main phenolic compounds isolated from hot pepper fruit (Capsicum annuum L.). *Journal of Agricultural Food Chemistry*, 53:1750.

Moron, M.S., Despierre, J.W., Minnervik, B.(1979). Levels of glutathione, glutathione reductase and glutathione-S-transferase activities in rat lung and liver. *Biochima et Biophysics Acta,* 582:67.

Ohkawa, H., Ohish, N., Yagi, K. (1979). Assay of lipid peroxides in animal tissues by thiobarbituric acid reaction. *Analytical Biochemistry,* 95:351.

Omaye, A.T., Turnbull, J.D., Sauberdich, H.E. (1971). Selected method for determination of ascorbic acid in animal cells, tissues and fluids. *Methods in Enzymology,*62:1.

Pushpavalli, G., Kalaiarasi, P., Veeramani, C., Pugalendi, K.V. (2010). Effect of chrysin on hepatoprotective and antioxidant status in D-galactosamine-induced hepatitis in rats. *European Journal of Pharmacology,* 10:36-41.

Ramakrishnan Gopalakrishnan., Raghavendran Hanumantha Rao Balaji., Vinodh Kumar Radhakrishnan., Devaki Thiruvengadam. (2006).

Suppression of N-nitrosodiethylamine induced hepatocarcinogenesis by silymarin in rats. *Chemico Biological interactions,* 161:104.

Ravikumar, V., Shivashangari, K.S., Devaki, T. (2005a). Hepatoprotective activity of Tridax procumbens against d-galactosamine/lipopolysaccharide -induced hepatitis in rats. *Journal of Ethnopharmacology*, 101(1-3):55-60.

Ravikumar, V., Shivashangari, K.S., Devaki, T. (2005b). Effect of Tridax procumbens on liver antioxidant defense system during lipopolysaccharide-induced hepatitis in D-galactosamine sensitised rats.Molecular and Cellular Biochemistry, 269(1-2):131-136.

Rice Evans, C.A., Miller, J., Paganga, G. (1997). Antioxidant properties of phenolic compounds. *Trends in Plant Science*, 2:152.

Rotruck, J.T., Pope, A.L., Ganther, H.E., Swanson, A.B., Hajeman, D.G., Hoekstra, W.G. (1973). Selenium-biochemical role as a component of glutathione purification and assay. *Science*,179:588.

Sathivel, A., Balavinayagamani., Hanumantha Rao,B.R., Devaki, T. (2013). Sulfated polysaccharide isolated from Ulva lactuca attenuates d-galactosamine induced DNA fragmentation and necrosis during liver damage in rats. *Pharmaceutical Biology,* (In Press).

Sathivel, A., Raghavendran, H.R., Srinivasan, P., Devaki, T. (2008). Anti-peroxidative and anti-hyperlipidemic nature of Ulva lactuca crude polysaccharide on D-galactosamine induced hepatitis in rats. *Food and Chemical Toxicology,* 46(10):3262-3267.

Serbecic, N., Ehmann, A.K., Beutelspacher, S.C. (2005). Reduction of lipid peroxidation and apoptosis in corneal endothelial cells by vitamin A. *Ophthalmologe,* 6:607.

Sinha, A.K. (1974). Colorimetric assay of catalase. *Analytical Biochemistry*, 47:389.

Sheik Abdulazeez, S., Thiruvengadam, D. (2013). Effect of lycopene on oxidative stress induced during D-galactosamine/lipopolysaccharide-sensitized liver injury in rats. *Pharmaceutical Biology*, 51(12):1592-1599.

Sriram, N., Kalayarasan, S., Sudhandiran, G (2008). Enhancement of antioxidant defense system by epigallocatechin-3-gallate during bleomycin induced experimental pulmonary fibrosis. *Biological Pharmaceutical Bulletin*, 31(7): 1306-1311.

Surh, Y.J., Lee, R.C., Park, K.K., Mayne, S.T., Liem, A., Miller, J.A. (1995). Chemopreventive effects of capsaicin and diallyl sulfide against mutagenesis or tumorigenesis by vinyl carbamate and N-nitrosodimethylamine. *Carcinogenesis,* 16:2467.

Wang, H., Nair, M.G., Strasburg, G.M., Chang, Y.C., Booren, A.M., Gray, J.L., Dewitt, D.L. (1999). Antioxidant and anti-inflammatory activities of anthocyanins and their aglycon, cyaniding from tart cherries. *Journal of Natural Products*, 62: 294-296.

Werawatganon, D., Linlawan, S., Thanapirom, K., Somanawat, K., Klaikeaw, N., Rerknimitr, R., Sriviriyakul, P. (2014). Aloe Vera attenuated liver injury in mice with acetaminophen-induced hepatitis. *BMC: Complementary and alternative medicine*. 14: 229.

INDEX

B

C

H

I

J

K

L

M

N

O

P

Q

R

S

T